M000250259

HOLT SCIENCE & TECHNOLOGY

Introduction to Science

HOLT, RINEHART AND WINSTON

A Harcourt Education Company

Orlando • Austin • New York • San Diego • Toronto • London

Acknowledgments

Contributing Authors

Katy Z. Allen
Science Writer
Wayland, Massachusetts

Leila Dumas
Former Physics Teacher
Austin, Texas

Robert H. Fronk, Ph.D.
Professor
Science and Mathematics
 Education Department
Florida Institute of
 Technology
Melbourne, Florida

Inclusion Specialist

Karen Clay
*Inclusion Specialist
 Consultant*
Boston, Massachusetts

Ellen McPeek Glisan
Special Needs Consultant
San Antonio, Texas

Safety Reviewer

Jack Gerlovich, Ph.D.
Associate Professor
School of Education
Drake University
Des Moines, Iowa

Academic Reviewers

Jim Denbow, Ph.D.
*Associate Professor of
 Archaeology*
Department of
 Anthropology
 and Archaeology
The University of Texas
 at Austin
Austin, Texas

Cassandra Eagle
Professor
Chemistry Department
Appalachian State
 University
Boone, North Carolina

Daniela Kohen
*Assistant Professor of
 Chemistry*
Chemistry Department
Carleton College
North Field, Minnesota

Steven A. Jennings, Ph.D.
Associate Professor
Geography and
 Environmental Studies
University of Colorado at
 Colorado Springs
Colorado Springs, Colorado

Joel S. Leventhal, Ph.D.
Emeritus Scientist
U.S. Geological Survey
Lakewood, Colorado

Laurie Santos, Ph.D.
Assistant Professor
Department of Psychology
Yale University
New Haven, Connecticut

Fred Seaman, Ph.D.
Retired Research Associate
College of Pharmacy
The University of Texas at
 Austin
Austin, Texas

Copyright © 2005 by Holt, Rinehart and Winston

All rights reserved. No part of this publication may be reproduced or transmitted in any form or by any means, electronic or mechanical, including photocopy, recording, or any information storage and retrieval system, without permission in writing from the publisher.

Requests for permission to make copies of any part of the work should be mailed to the following address: Permissions Department, Holt, Rinehart and Winston, 10801 N. MoPac Expressway, Building 3, Austin, Texas 78759.

SciLINKS is a registered trademark owned and provided by the National Science Teachers Association. All rights reserved.

CNN is a registered trademark of Cable News Network LP, LLLP, an AOL Time Warner Company.

Current Science® is a registered trademark of Weekly Reader Corporation.

Printed in the United States of America

ISBN 0-03-042398-8

1 2 3 4 5 6 7 048 09 08 07 06 05 04

Acknowledgments

Teacher Reviewers

Robin K. Clanton
Science Department Head
Berrien Middle School
Nashville, Georgia

Randy Dye, M.S.
Middle School Science Department Head
Earth Science
Wood Middle School
Waynesville School District #6, Missouri

Jack Gerlovich, Ph.D.
Associate Professor
School of Education
Drake University
Des Moines, Iowa

Ronald W. Hudson
Science Teacher
Batchelor Middle School
Bloomington, Indiana

Lab Testing

Paul Boyle
Science Teacher
Perry Heights Middle School
Evansville, Indiana

Georgiann Delgadillo
Science Teacher
East Valley Continuous Curriculum School
Spokane, Washington

Edith C. McAlanis
Science Teacher and Department Chair
Socorro Middle School
El Paso, Texas

Terry J. Rakes
Science Teacher
Elmwood Junior High School
Rogers, Arkansas

Larry Tackett
Science Teacher and Dept. Chair
R. H. Terrell Junior High School
Washington, D.C.

Feature Development

Katy Z. Allen
Mickey Coakley
Jane Gardner
Catherine Podeszwa

Answer Checking

Hatim Belyamani
Austin, Texas
John A. Benner
Austin, Texas
Catherine Podeszwa
Duluth, Minnesota

Introduction to Science

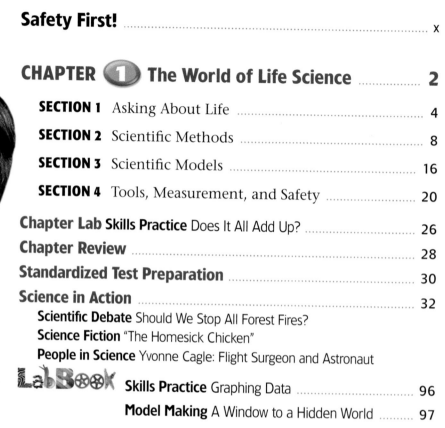

Skills Development

Connection to . . .

Science in Action

How to Use Your Textbook

Your Roadmap for Success with Holt Science and Technology

Reading Warm-Up

A Reading Warm-Up at the beginning of every section provides you with the section's objectives and key terms. The objectives tell you what you'll need to know after you finish reading the section.

Key terms are listed for each section. Learn the definitions of these terms because you will most likely be tested on them. Each key term is highlighted in the text and is defined at point of use and in the margin. You can also use the glossary to locate definitions quickly.

STUDY TIP Reread the objectives and the definitions to the key terms when studying for a test to be sure you know the material.

Get Organized

A Reading Strategy at the beginning of every section provides tips to help you organize and remember the information covered in the section. Keep a science notebook so that you are ready to take notes when your teacher reviews the material in class. Keep your assignments in this notebook so that you can review them when studying for the chapter test.

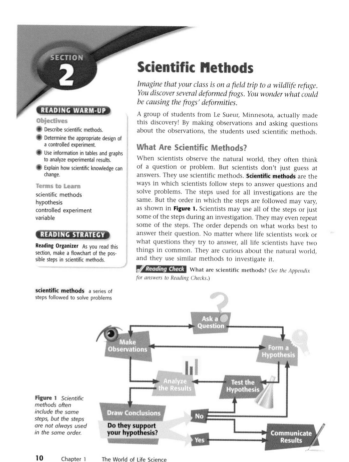

SECTION 2

Scientific Methods

Imagine that your class is on a field trip to a wildlife refuge. You discover several deformed frogs. You wonder what could be causing the frogs' deformities.

A group of students from Le Sueur, Minnesota, actually made this discovery! By making observations and asking questions about the observations, the students used scientific methods.

READING WARM-UP

Objectives
● Describe scientific methods.
● Determine the appropriate design of a controlled experiment.
● Use information in tables and graphs to analyze experimental results.
● Explain how scientific knowledge can change.

Terms to Learn
scientific methods
hypothesis
controlled experiment
variable

READING STRATEGY

Reading Organizer As you read this section, make a flowchart of the possible steps in scientific methods.

scientific methods a series of steps followed to solve problems

What Are Scientific Methods?

When scientists observe the natural world, they often think of a question or problem. But scientists don't just guess at answers. They use scientific methods. **Scientific methods** are the ways in which scientists follow steps to answer questions and solve problems. The steps used for all investigations are the same. But the order in which the steps are followed may vary, as shown in **Figure 1**. Scientists may use all of the steps or just some of the steps during an investigation. They may even repeat some of the steps. The order depends on what works best to answer their question. No matter where life scientists work or what questions they try to answer, all life scientists have two things in common. They are curious about the natural world, and they use similar methods to investigate it.

Reading Check What are scientific methods? *(See the Appendix for answers to Reading Checks.)*

Figure 1 *Scientific methods often include the same steps, but the steps are not always used in the same order.*

10 Chapter 1 The World of Life Science

Be Resourceful—Use the Web

SciLinks boxes in your textbook take you to resources that you can use for science projects, reports, and research papers. Go to **scilinks.org** and type in the **SciLinks code** to find information on a topic.

Visit go.hrw.com
Find worksheets, **Current Science**® magazine articles online, and other materials that go with your textbook at **go.hrw.com.** Click on the textbook icon and the table of contents to see all of the resources for each chapter.

Ask a Question

Have you ever observed something out of the ordinary or difficult to explain? Such an observation usually raises questions. For example, you might ask, "Could something in the water be causing the frog deformities?" Looking for answers may include making more observations.

Make Observations

After the students from Minnesota realized something was wrong with the frogs, they decided to make additional, careful observations, as shown in **Figure 2.** They counted the number of deformed frogs and the number of normal frogs they caught. The students also photographed the frogs, took measurements, and wrote a thorough description of each frog.

In addition, the students collected data on other organisms living in the pond. They also conducted many tests on the pond water, measuring things such as the level of acidity. The students carefully recorded their data and observations.

Accurate Observations

Any information you gather through your senses is an observation. Observations can take many forms. They may be measurements of length, volume, time, or speed or of how loud or soft a sound is. They may describe the color or shape of an organism. Or they may record the behavior of organisms in an area. The range of observations a scientist can make is endless. But no matter what observations reveal, they are useful only if they are accurately made and recorded. Scientists use many standard tools and methods to make and record observations. Examples of these tools are shown in **Figure 3.**

Figure 2 *Making careful observations is often the first step in an investigation.*

Figure 3 *Microscopes, rulers, and thermometers are some of the many tools scientists use to collect information. Scientists also record their observations carefully.*

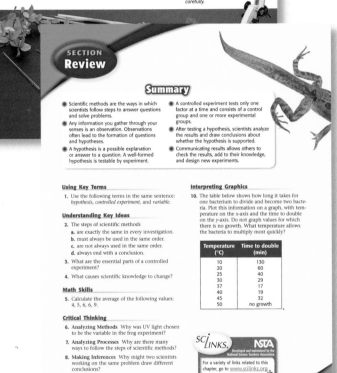

SECTION Review

Summary

- Scientific methods are the ways in which scientists follow steps to answer questions and solve problems.
- Any information you gather through your senses is an observation. Observations often lead to the formation of questions and hypotheses.
- A hypothesis is a possible explanation or answer to a question. A well-formed hypothesis is testable by experiment.
- A controlled experiment tests only one factor at a time and consists of a control group and one or more experimental groups.
- After testing a hypothesis, scientists analyze the results and draw conclusions about whether the hypothesis is supported.
- Communicating results allows others to check the results, add to their knowledge, and design new experiments.

Using Key Terms

1. Use the following terms in the same sentence: *hypothesis, controlled experiment,* and *variable.*

Understanding Key Ideas

2. The steps of scientific methods
   a. are exactly the same in every investigation.
   b. must always be used in the same order.
   c. are not always used in the same order.
   d. always end with a conclusion.

3. What are the essential parts of a controlled experiment?

4. What causes scientific knowledge to change?

Math Skills

5. Calculate the average of the following values: 4, 5, 6, 6, 9.

Critical Thinking

6. **Analyzing Methods** Why was UV light chosen to be the variable in the frog experiment?

7. **Analyzing Processes** Why are there many ways to follow the steps of scientific methods?

8. **Making Inferences** Why might two scientists working on the same problem draw different conclusions?

9. **Identifying Bias** Investigations often begin with observation. How does observation limit what scientists can study?

Interpreting Graphics

10. The table below shows how long it takes for one bacterium to divide and become two bacteria. Plot this information on a graph, with temperature on the *x*-axis and the time to double on the *y*-axis. Do not graph values for which there is no growth. What temperature allows the bacteria to multiply most quickly?

Temperature (°C)	Time to double (min)
10	130
20	60
25	40
30	29
37	17
40	19
45	32
50	no growth

SCLINKS. Developed and maintained by the National Science Teachers Association

For a variety of links related to this chapter, go to www.scilinks.org
Topic: Scientific Methods; Deformed Frogs
SciLinks code: HSM1359; HSM0583

17

Use the Illustrations and Photos

Art shows complex ideas and processes. Learn to analyze the art so that you better understand the material you read in the text.

Tables and graphs display important information in an organized way to help you see relationships.

A picture is worth a thousand words. Look at the photographs to see relevant examples of science concepts that you are reading about.

Answer the Section Reviews

Section Reviews test your knowledge of the main points of the section. Critical Thinking items challenge you to think about the material in greater depth and to find connections that you infer from the text.

STUDY TIP When you can't answer a question, reread the section. The answer is usually there.

Do Your Homework

Your teacher may assign worksheets to help you understand and remember the material in the chapter.

STUDY TIP Don't try to answer the questions without reading the text and reviewing your class notes. A little preparation up front will make your homework assignments a lot easier. Answering the items in the Chapter Review will help prepare you for the chapter test.

Holt Online Learning

Visit Holt Online Learning

If your teacher gives you a special password to log onto the **Holt Online Learning** site, you'll find your complete textbook on the Web. In addition, you'll find some great learning tools and practice quizzes. You'll be able to see how well you know the material from your textbook.

SAFETY FIRST!

Exploring, inventing, and investigating are essential to the study of science. However, these activities can also be dangerous. To make sure that your experiments and explorations are safe, you must be aware of a variety of safety guidelines. You have probably heard of the saying, "It is better to be safe than sorry." This is particularly true in a science classroom where experiments and explorations are being performed. Being uninformed and careless can result in serious injuries. Don't take chances with your own safety or with anyone else's.

The following pages describe important guidelines for staying safe in the science classroom. Your teacher may also have safety guidelines and tips that are specific to your classroom and laboratory. Take the time to be safe.

Safety Rules!

Start Out Right

Always get your teacher's permission before attempting any laboratory exploration. Read the procedures carefully, and pay particular attention to safety information and caution statements. If you are unsure about what a safety symbol means, look it up or ask your teacher. You cannot be too careful when it comes to safety. If an accident does occur, inform your teacher immediately regardless of how minor you think the accident is.

Safety Symbols

All of the experiments and investigations in this book and their related worksheets include important safety symbols to alert you to particular safety concerns. Become familiar with these symbols so that when you see them, you will know what they mean and what to do. It is important that you read this entire safety section to learn about specific dangers in the laboratory.

If you are instructed to note the odor of a substance, wave the fumes toward your nose with your hand. Never put your nose close to the source.

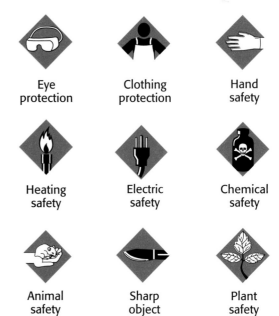

Eye protection

Clothing protection

Hand safety

Heating safety

Electric safety

Chemical safety

Animal safety

Sharp object

Plant safety

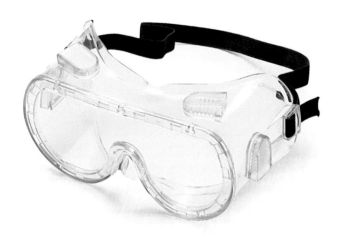

Eye Safety

Wear safety goggles when working around chemicals, acids, bases, or any type of flame or heating device. Wear safety goggles any time there is even the slightest chance that harm could come to your eyes. If any substance gets into your eyes, notify your teacher immediately and flush your eyes with running water for at least 15 minutes. Treat any unknown chemical as if it were a dangerous chemical. Never look directly into the sun. Doing so could cause permanent blindness.

Avoid wearing contact lenses in a laboratory situation. Even if you are wearing safety goggles, chemicals can get between the contact lenses and your eyes. If your doctor requires that you wear contact lenses instead of glasses, wear eye-cup safety goggles in the lab.

Safety Equipment

Know the locations of the nearest fire alarms and any other safety equipment, such as fire blankets and eyewash fountains, as identified by your teacher, and know the procedures for using the equipment.

Neatness

Keep your work area free of all unnecessary books and papers. Tie back long hair, and secure loose sleeves or other loose articles of clothing, such as ties and bows. Remove dangling jewelry. Don't wear open-toed shoes or sandals in the laboratory. Never eat, drink, or apply cosmetics in a laboratory setting. Food, drink, and cosmetics can easily become contaminated with dangerous materials.

Certain hair products (such as aerosol hair spray) are flammable and should not be worn while working near an open flame. Avoid wearing hair spray or hair gel on lab days.

Sharp/Pointed Objects

Use knives and other sharp instruments with extreme care. Never cut objects while holding them in your hands. Place objects on a suitable work surface for cutting.

Be extra careful when using any glassware. When adding a heavy object to a graduated cylinder, tilt the cylinder so the object slides slowly to the bottom.

Heat

Wear safety goggles when using a heating device or a flame. Whenever possible, use an electric hot plate as a heat source instead of using an open flame. When heating materials in a test tube, always angle the test tube away from yourself and others. To avoid burns, wear heat-resistant gloves whenever instructed to do so.

Electricity

Be careful with electrical cords. When using a microscope with a lamp, do not place the cord where it could trip someone. Do not let cords hang over a table edge in a way that could cause equipment to fall if the cord is accidentally pulled. Do not use equipment with damaged cords. Be sure that your hands are dry and that the electrical equipment is in the "off" position before plugging it in. Turn off and unplug electrical equipment when you are finished.

Chemicals

Wear safety goggles when handling any potentially dangerous chemicals, acids, or bases. If a chemical is unknown, handle it as you would a dangerous chemical. Wear an apron and protective gloves when you work with acids or bases or whenever you are told to do so. If a spill gets on your skin or clothing, rinse it off immediately with water for at least 5 minutes while calling to your teacher.

Never mix chemicals unless your teacher tells you to do so. Never taste, touch, or smell chemicals unless you are specifically directed to do so. Before working with a flammable liquid or gas, check for the presence of any source of flame, spark, or heat.

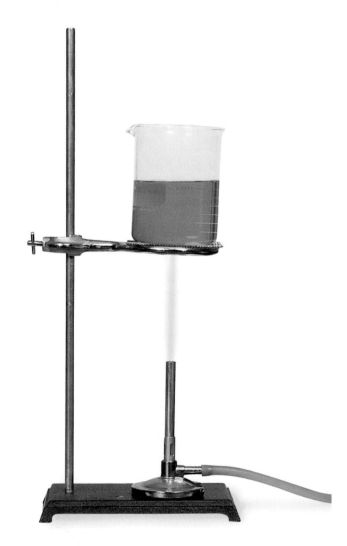

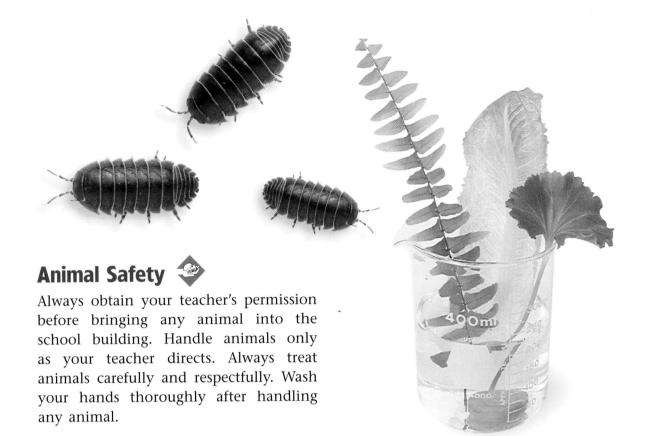

Animal Safety

Always obtain your teacher's permission before bringing any animal into the school building. Handle animals only as your teacher directs. Always treat animals carefully and respectfully. Wash your hands thoroughly after handling any animal.

Plant Safety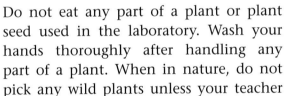

Do not eat any part of a plant or plant seed used in the laboratory. Wash your hands thoroughly after handling any part of a plant. When in nature, do not pick any wild plants unless your teacher instructs you to do so.

Glassware

Examine all glassware before use. Be sure that glassware is clean and free of chips and cracks. Report damaged glassware to your teacher. Glass containers used for heating should be made of heat-resistant glass.

1

The World of Life Science

About the

What happened to the legs of these frogs? Life science can help answer this question. Deformed frogs, such as the ones in this photo, have been found in the northern United States and southern Canada. Scientists and students like you have been using life science to find out how frogs can develop deformities.

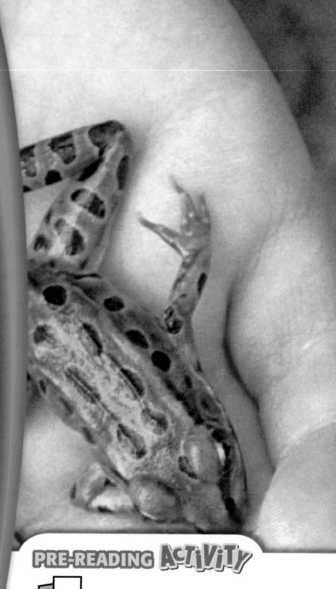

PRE-READING ACTIVITY

FOLDNOTES **Layered Book** Before you read the chapter, create the FoldNote entitled "Layered Book" described in the **Study Skills** section of the Appendix. Label the tabs of the layered book with "Examples of life scientists," "Scientific methods," "Scientific models," and "Tools, measurement, and safety." As you read the chapter, write information you learn about each category under the appropriate tab.

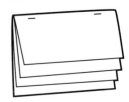

START-UP ACTIVITY

A Little Bit of Science

In this activity, you'll find out that you can learn about the unknown without having to see it.

Procedure

1. Your teacher will give you a **coffee can** to which a **sock** has been attached. Do not look into the can.

2. Reach through the opening in the sock. You will feel **several objects** inside the can.

3. Record observations you make about the objects by feeling them, shaking the can, and so on.

4. What do you think is in the can? List your guesses. State some reasons for your guesses.

5. Pour the contents of the can onto your desk. Compare your list with what was in the can.

Analysis

1. Did you guess the contents of the can correctly? What might have caused you to guess wrongly?

2. What observations did you make about each of the objects while they were in the can? Which of your senses did you use?

Asking About Life

Imagine that it's summer. You are lying in the grass at the park, casually looking around. Three dogs are playing on your left. A few bumblebees are visiting nearby flowers. And an ant is carrying a crumb away from your sandwich.

Suddenly, a question pops into your head: How do ants find food? Then, you think of another question: Why do the bees visit the yellow flowers but not the red ones? Congratulations! You have just taken the first steps toward becoming a life scientist. How did you do it? You observed the living world around you. You were curious, and you asked questions about your observations. Those steps are what science is all about. **Life science** is the study of living things.

✓ **Reading Check** What is life science? (*See the Appendix for answers to Reading Checks.*)

It All Starts with a Question

The world around you is full of an amazing diversity of life. Single-celled algae float unseen in ponds. Giant redwood trees seem to touch the sky. And 40-ton whales swim through the oceans. For every living thing, or organism, that has ever lived, you could ask many questions. Those questions could include (1) How does the organism get its food? (2) Where does it live? and (3) Why does it behave in a particular way?

In Your Own Backyard

Questions are easy to think of. Take a look around your room, your home, and your neighborhood. What questions about life science come to mind? The student in **Figure 1** has questions about some very familiar organisms. Do you know the answer to any of his questions?

Touring the World

The questions you can ask about your neighborhood are just a sample of all the questions you could ask about the world. The world is made up of many different places to live, such as deserts, forests, coral reefs, and tide pools. Just about anywhere you go, you will find some kind of living organism. If you observe these organisms, you can easily think of questions to ask about them.

READING WARM-UP

Objectives

● Explain the importance of asking questions in life science.

● State examples of life scientists at work.

● List three ways life science is beneficial to living things.

Terms to Learn

life science

READING STRATEGY

Paired Summarizing Read this section silently. In pairs, take turns summarizing the material. Stop to discuss ideas that seem confusing.

life science the study of living things

Why do leaves change color in the fall?

Why did the dinosaurs die out?

How do birds know where to go when they migrate?

Figure 1 *Part of science is asking questions about the world around you.*

4

Irene Duhart Long asks, "How does the human body respond to space travel?"

Geerat Vermeij asks, "How have shells changed over time?"

Irene Pepperberg asks, "Are parrots smart enough to learn human language?"

Figure 2 *Life scientists ask many different kinds of questions about living things.*

Life Scientists

Close your eyes for a moment, and imagine a life scientist. What do you see? Do you see someone who is in a laboratory and peering into a microscope? Which of the people in **Figure 2** do you think are life scientists?

Anyone

If you guessed that all of the people in **Figure 2** are life scientists, then you are right. Anyone can investigate the world around us. Women and men from any cultural or ethnic background can become life scientists.

Anywhere

Making investigations in a laboratory is an important part of life science, but life science can be studied in many other places, too. Life scientists carry out investigations on farms, in forests, on the ocean floor—even in space. They work for businesses, hospitals, government agencies, and universities. Many are also teachers.

Anything

What a life scientist studies is determined by one thing—his or her curiosity. Life scientists specialize in many different areas of life science. They may study how organisms function and behave. Or they may study how organisms interact with each other and with their environment. Some life scientists explore how organisms reproduce and pass traits from one generation to the next. Some life scientists investigate the ancient origins of organisms and the ways in which organisms have changed over time.

CONNECTION TO Language Arts

WRITING SKILL **Profile of a Life Scientist** Research some of the life scientists named in this chapter. Choose the scientist who interests you the most. In your **science journal,** write a short biography, career feature, or informational piece about your chosen scientist and the work he or she does. Style the article as a newspaper or magazine article.

How do certain chemicals affect the virus that causes AIDS?

Figure 3 *Abdul Lakhani studies AIDS to try to find a cure for the disease.*

Why Ask Questions?

What is the point of asking all these questions? Life scientists might find some interesting answers, but do any of the answers really matter? Will the answers affect *your* life? Absolutely! As you study life science, you will see how the investigations of life science affect you and all the living things around you.

Fighting Diseases

Polio is a disease that causes paralysis by affecting the brain and nerves. Do you know anyone who has had polio? Probably not. The polio virus has been eliminated from most of the world. But at one time, it was much more common. In 1952, before life scientists discovered ways to prevent the spread of the polio virus, it infected 58,000 Americans.

Today, life scientists continue to search for ways to fight diseases. Acquired immune deficiency syndrome (AIDS) is a disease that kills millions of people every year. The scientist in **Figure 3** is trying to learn more about AIDS. Life scientists have discovered how the virus that causes AIDS is carried from one person to another. Scientists have also learned about how the virus affects the body. By learning more about the virus, scientists may find a cure for this deadly disease.

Understanding Inherited Diseases

Some diseases, such as cystic fibrosis, are inherited. They are passed from parents to children. Most of the information that controls an organism's cells is inherited as coded information. Changes in small parts of this information may cause the organism to be born with or to develop certain diseases. The scientist in **Figure 4** is one of the many scientists worldwide who are studying the way humans inherit the code that controls their cells. By learning about this code, scientists hope to find ways to cure or prevent inherited diseases.

Figure 4 *Susumu Tonegawa's work may help in the battle to fight inherited diseases.*

Which part of a person's inherited information is responsible for certain inherited diseases?

Protecting the Environment

Life scientists also study environmental problems on Earth. Many environmental problems are caused by people's misuse of natural resources. Understanding how we affect the world around us is the first step in finding solutions to problems such as pollution and the extinction of wildlife.

Why should we try to decrease pollution? Pollution can harm our health and the health of other organisms. Water pollution may be a cause of frog deformities seen in Minnesota and other states. Pollution in oceans kills marine mammals, birds, and fish. By finding ways to produce less pollution, we can help make the world a healthier place.

When we cut down trees to clear land for crops or for lumber, we change and sometimes destroy habitats. The man in **Figure 5** is part of a team of Russian and American scientists studying the Siberian tiger. Hunting and loss of forests have caused the tigers to become almost extinct. By learning about the tigers' food and habitat needs, the scientists hope to develop a plan that will ensure their survival.

> How much space does a tiger need in order to survive?

Figure 5 *To learn how much land area is used by an individual Siberian tiger, Dale Miquelle puts radio-transmitting collars on the tigers.*

✔ Reading Check Give an example of a pollution problem.

SECTION Review

Summary

● Science is a process of gathering knowledge about the natural world. Science includes making observations and asking questions about those observations. Life science is the study of living things.

● A variety of people may become life scientists for a variety of reasons.

● Life science can help solve problems such as disease or pollution, and it can be applied to help living things survive.

Using Key Terms

1. In your own words, write a definition for the term *life science*.

Understanding Key Ideas

2. Life scientists may study any of the following EXCEPT
   a. things that were once living.
   b. environmental problems.
   c. stars in outer space.
   d. diseases that are not inherited by humans.

3. What is the importance of asking questions in life science?

4. Where do life scientists work? What do life scientists study?

Math Skills

5. Students in a science class collected 50 frogs from a pond and found that 15 of these frogs had deformities. What percentage of the frogs had deformities?

Critical Thinking

6. **Identifying Relationships** Make a list of five things you do or deal with daily. Give an example of how life science might relate to each of these things.

7. **Applying Concepts** Look at **Figure 5.** Propose five questions about what you see. Share one of your questions with your classmates.

SCiLINKS®

NSTA
Developed and maintained by the
National Science Teachers Association

For a variety of links related to this chapter, go to www.scilinks.org

Topic: Careers in Life Science
SciLinks code: HSM0224

Scientific Methods

Imagine that your class is on a field trip to a wildlife refuge. You discover several deformed frogs. You wonder what could be causing the frogs' deformities.

A group of students from Le Sueur, Minnesota, actually made this discovery! By making observations and asking questions about the observations, the students used scientific methods.

What Are Scientific Methods?

When scientists observe the natural world, they often think of a question or problem. But scientists don't just guess at answers. They use scientific methods. **Scientific methods** are the ways in which scientists follow steps to answer questions and solve problems. The steps used for all investigations are the same. But the order in which the steps are followed may vary, as shown in **Figure 1.** Scientists may use all of the steps or just some of the steps during an investigation. They may even repeat some of the steps. The order depends on what works best to answer their question. No matter where life scientists work or what questions they try to answer, all life scientists have two things in common. They are curious about the natural world, and they use similar methods to investigate it.

✓ **Reading Check** What are scientific methods? (*See the Appendix for answers to Reading Checks.*)

READING WARM-UP

Objectives

● Describe scientific methods.

● Determine the appropriate design of a controlled experiment.

● Use information in tables and graphs to analyze experimental results.

● Explain how scientific knowledge can change.

Terms to Learn

scientific methods
hypothesis
controlled experiment
variable

READING STRATEGY

Reading Organizer As you read this section, make a flowchart of the possible steps in scientific methods.

scientific methods a series of steps followed to solve problems

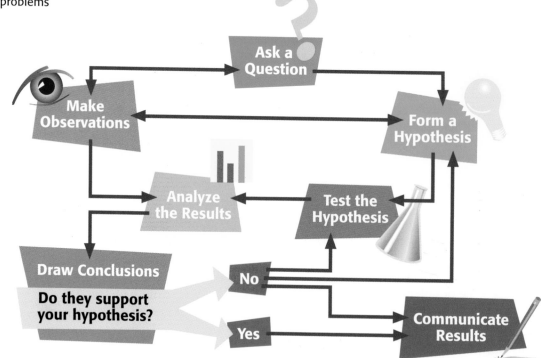

Figure 1 *Scientific methods often include the same steps, but the steps are not always used in the same order.*

Ask a Question

Have you ever observed something out of the ordinary or difficult to explain? Such an observation usually raises questions. For example, you might ask, "Could something in the water be causing the frog deformities?" Looking for answers may include making more observations.

Make Observations

After the students from Minnesota realized something was wrong with the frogs, they decided to make additional, careful observations, as shown in **Figure 2.** They counted the number of deformed frogs and the number of normal frogs they caught. The students also photographed the frogs, took measurements, and wrote a thorough description of each frog.

In addition, the students collected data on other organisms living in the pond. They also conducted many tests on the pond water, measuring things such as the level of acidity. The students carefully recorded their data and observations.

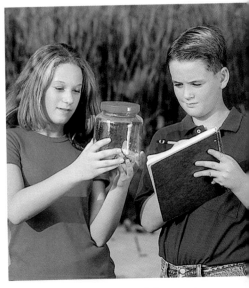

Figure 2 *Making careful observations is often the first step in an investigation.*

Accurate Observations

Any information you gather through your senses is an observation. Observations can take many forms. They may be measurements of length, volume, time, or speed or of how loud or soft a sound is. They may describe the color or shape of an organism. Or they may record the behavior of organisms in an area. The range of observations a scientist can make is endless. But no matter what observations reveal, they are useful only if they are accurately made and recorded. Scientists use many standard tools and methods to make and record observations. Examples of these tools are shown in **Figure 3.**

Figure 3 *Microscopes, rulers, and thermometers are some of the many tools scientists use to collect information. Scientists also record their observations carefully.*

CONNECTION TO
Environmental Science

WRITING
SKILL   Vanishing
         Amphibians

Since the 1980s, scientists
have been concerned about
a steady worldwide decline
in the number of amphibians,
such as frogs and salamanders.
Scientists have studied several
possible causes, including UV
radiation, chemical pollutants,
parasites, and skin fungi. Find
a recent news article about
one such study, and write a
short summary of the article.

Form a Hypothesis

After asking questions and making observations, scientists may form a hypothesis. A **hypothesis** (hie PAHTH uh sis) is a possible explanation or answer to a question. A good hypothesis is based on observation and can be tested. When scientists form hypotheses, they think logically and creatively and consider what they already know.

To be useful, a hypothesis must be testable. A hypothesis is testable if an experiment can be designed to test the hypothesis. Yet, if a hypothesis is not testable, it is not always wrong. An untestable hypothesis is simply one that cannot be supported or disproved. Sometimes, it may be impossible to gather enough observations to test a hypothesis.

Scientists may form different hypotheses for the same problem. In the case of the Minnesota frogs, scientists formed the hypotheses shown in **Figure 4.** Were any of these explanations correct? To find out, each hypothesis had to be tested.

✓ **Reading Check** What makes a hypothesis testable?

hypothesis an explanation that is based on prior scientific research or observations and that can be tested

Figure 4
More than one hypothesis can be made for a single question.

Hypothesis 1:
The deformities were caused by one or more chemical pollutants in the water.

Hypothesis 2:
The deformities were caused by attacks from parasites or other frogs.

Hypothesis 3:
The deformities were caused by an increase in exposure to ultraviolet light from the sun.

Predictions

Before scientists can test a hypothesis, they must first make predictions. A prediction is a statement of cause and effect that can be used to set up a test for a hypothesis. Predictions are usually stated in an if-then format, as shown in **Figure 5.**

More than one prediction may be made for each hypothesis. For each of the hypotheses on the previous page, the predictions shown in **Figure 5** were made. After predictions are made, scientists can conduct experiments to see which predictions, if any, prove to be true and support the hypotheses.

Figure 5 *More than one prediction may be made for a single hypothesis.*

Hypothesis 1:
Prediction: *If* a substance in the pond water is causing the deformities, *then* the water from ponds that have deformed frogs will be different from the water from ponds in which no abnormal frogs have been found.
Prediction: *If* a substance in the pond water is causing the deformities, *then* some tadpoles will develop deformities when they are raised in pond water collected from ponds that have deformed frogs.

Hypothesis 2:
Prediction: *If* a parasite is causing the deformities, *then* this parasite will be found more often in frogs that have deformities.

Hypothesis 3:
Prediction: *If* an increase in exposure to ultraviolet light is causing the deformities, *then* some frog eggs exposed to ultraviolet light in a laboratory will develop into deformed frogs.

CONNECTION TO
Language Arts

WRITING SKILL **Have Aliens Landed?** Suppose that you and a friend are walking through a heavily wooded park. Suddenly, you come upon a small cluster of trees lying on the ground. What caused them to fall over? Your friend thinks that extraterrestrials knocked the trees down. Write a dialogue of the debate you might have with your friend about whether this hypothesis is testable.

Test the Hypothesis

After scientists make a prediction, they test the hypothesis. Scientists try to design experiments that will clearly show whether a particular factor caused an observed outcome. A *factor* is anything in an experiment that can influence the experiment's outcome. Factors can be anything from temperature to the type of organism being studied.

Under Control

Scientists studying the frogs in Minnesota observed many factors that affect the development of frogs in the wild, as shown in **Figure 6.** But it was hard to tell which factor could be causing the deformities. To sort factors out, scientists perform controlled experiments. A **controlled experiment** tests only one factor at a time and consists of a control group and one or more experimental groups. All of the factors for the control group and the experimental groups are the same except for one. The one factor that differs is called the **variable.** Because only the variable differs between the control group and the experimental groups, any differences observed in the outcome of the experiment are probably caused by the variable.

✓ **Reading Check** How many factors should an experiment test?

Designing an Experiment

Designing a good experiment requires planning. Every factor should be considered. Examine the prediction for Hypothesis 3: *If an increase in exposure to ultraviolet light is causing the deformities, then some frog eggs exposed to ultraviolet light in a laboratory will develop into deformed frogs.* An experiment to test this hypothesis is summarized in **Table 1.** In this case, the variable is the length of time the eggs are exposed to ultraviolet (UV) light. All other factors, such as the temperature of the water, are the same in the control group and in the experimental groups.

Figure 6 *Many factors affect this tadpole in the wild. These factors include chemicals, light, temperature, and parasites.*

controlled experiment an experiment that tests only one factor at a time by using a comparison of a control group with an experimental group

variable a factor that changes in an experiment in order to test a hypothesis

Table 1 Experiment to Test Effect of UV Light on Frogs				
	Control factors			**Variable**
Group	**Kind of frog**	**Number of Eggs**	**Temperature of water**	**UV light exposure**
#1 (control)	leopard frog	100	25°C	0 days
#2 (experimental)	leopard frog	100	25°C	15 days
#3 (experimental)	leopard frog	100	25°C	24 days

Figure 7 UV Light Experiment

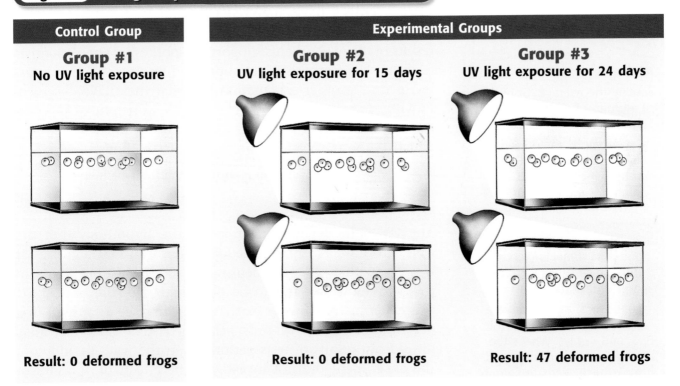

Control Group	Experimental Groups	
Group #1 No UV light exposure	**Group #2** UV light exposure for 15 days	**Group #3** UV light exposure for 24 days

Result: 0 deformed frogs   Result: 0 deformed frogs   Result: 47 deformed frogs

Collecting Data

As **Table 1** shows, each group in the experiment contains 100 eggs. Scientists always try to test many individuals. The more organisms tested, the more certain scientists can be of the data they collect in an experiment. They want to be certain that differences between control and experimental groups are actually caused by differences in the variable and not by any differences among the individuals. Scientists also support their conclusions by repeating their experiments. If an experiment produces the same results again and again, scientists can be more certain about the effect the variable has on the outcome of the experiment. The experimental setup to test Hypothesis 3 is illustrated in **Figure 7.** The results are also shown.

Analyze the Results

A scientist's work does not end when an experiment is finished. After scientists finish their tests, they must analyze the results. Scientists must organize the data so that they can be analyzed. For example, scientists may organize the data in a table or a graph. The data collected from the UV light experiment are shown in the bar graph in **Figure 8.** Analyzing results helps scientists explain and focus on the effect of the variable. For example, the graph shows that the length of UV exposure has an effect on the development of frog deformities.

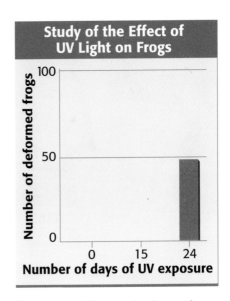

Figure 8 *This graph shows that 24 days of UV exposure had an effect on frog deformities, while less exposure had no effect.*

MATH PRACTICE

Averages

Finding the average, or mean, of a group of numbers is a common way to analyze data.

For example, three seeds were kept at 25°C and sprouted in 8, 8, and 5 days. To find the average number of days that it took the seeds to sprout, add 8, 8, and 5 and divide the sum by 3, the number of subjects (seeds). It took these seeds an average of 7 days to sprout.

Suppose three seeds were kept at 30°C and sprouted in 6, 5, and 4 days. What's the average number of days that it took these seeds to sprout?

Draw Conclusions

After scientists have analyzed the data from several experiments, they can draw conclusions. They decide whether the results of the experiments support a hypothesis. When scientists find that a hypothesis is not supported by the tests, they must try to find another explanation for what they have observed. Proving that a hypothesis is wrong is just as helpful as supporting it. Why? Either way, the scientist has learned something, which is the purpose of using scientific methods.

✔ **Reading Check** How can a wrong hypothesis be helpful?

Is It the Answer?

The UV light experiment supports the hypothesis that the frog deformities can be caused by exposure to UV light. Does this mean that UV light definitely caused the frogs living in the Minnesota wetland to be deformed? No, the only thing this experiment shows is that UV light may be a cause of frog deformities. Results of tests performed in a laboratory may differ from results of tests performed in the wild. In addition, the experiment did not investigate the effects of parasites or some other substance on the frogs. In fact, many scientists now think that more than one factor could be causing the deformities.

Puzzles as complex as the deformed-frog mystery are rarely solved with a single experiment. The quest for a solution may continue for years. Finding an answer doesn't always end an investigation. Often, that answer begins another investigation. In this way, scientists continue to build knowledge.

Communicate Results

Scientists form a global community. After scientists complete their investigations, they communicate their results to other scientists. The student in **Figure 9** is explaining the results of a science project.

There are several reasons scientists regularly share their results. First, other scientists may then repeat the experiments to see if they get the same results. Second, the information can be considered by other scientists with similar interests. The scientists can then compare hypotheses and form consistent explanations. New data may strengthen existing hypotheses or show that the hypotheses need to be altered. There are many paths from observations and questions to communicating results.

Figure 9 *This student scientist is communicating the results of his investigation at a science fair.*

SECTION Review

Summary

- Scientific methods are the ways in which scientists follow steps to answer questions and solve problems.

- Any information you gather through your senses is an observation. Observations often lead to the formation of questions and hypotheses.

- A hypothesis is a possible explanation or answer to a question. A well-formed hypothesis is testable by experiment.

- A controlled experiment tests only one factor at a time and consists of a control group and one or more experimental groups.

- After testing a hypothesis, scientists analyze the results and draw conclusions about whether the hypothesis is supported.

- Communicating results allows others to check the results, add to their knowledge, and design new experiments.

Using Key Terms

1. Use the following terms in the same sentence: *hypothesis, controlled experiment,* and *variable.*

Understanding Key Ideas

2. The steps of scientific methods

    a. are exactly the same in every investigation.

    b. must always be used in the same order.

    c. are not always used in the same order.

    d. always end with a conclusion.

3. What are the essential parts of a controlled experiment?

4. What causes scientific knowledge to change?

Math Skills

5. Calculate the average of the following values: 4, 5, 6, 6, 9.

Critical Thinking

6. **Analyzing Methods** Why was UV light chosen to be the variable in the frog experiment?

7. **Analyzing Processes** Why are there many ways to follow the steps of scientific methods?

8. **Making Inferences** Why might two scientists working on the same problem draw different conclusions?

9. **Identifying Bias** Investigations often begin with observation. How does observation limit what scientists can study?

Interpreting Graphics

10. The table below shows how long it takes for one bacterium to divide and become two bacteria. Plot this information on a graph, with temperature on the *x*-axis and the time to double on the *y*-axis. Do not graph values for which there is no growth. What temperature allows the bacteria to multiply most quickly?

Temperature (°C)	Time to double (min)
10	130
20	60
25	40
30	29
37	17
40	19
45	32
50	no growth

SCLINKS®

NSTA

Developed and maintained by the National Science Teachers Association

For a variety of links related to this chapter, go to www.scilinks.org

Topic: Scientific Methods; Deformed Frogs

SciLinks code: HSM1359; HSM0383

Scientific Models

How can you see the parts of a cell? Unless you had superhuman eyesight, you couldn't see inside most cells without a microscope.

How do you learn about the parts of the cell if you don't have a microscope? You can look at a model of a cell. A model can help you understand what the parts of a cell look like.

Types of Scientific Models

A **model** is a representation of an object or a system. Models are used in science to help explain how something works or to describe how something is structured. Models can also be used to make predictions or to explain observations. However, models have limitations. A model is never exactly like the real thing—if it were, it would no longer be a model. There are many kinds of scientific models. Some examples are physical models, mathematical models, and conceptual models.

Physical Models

A toy rocket and a plastic skeleton are examples of physical models. Many physical models, such as the model of a human body in **Figure 1,** look like the thing they model. However, a limitation of the model of a body is that it is not alive and doesn't act exactly like a human body. But the model is useful for understanding how the body works. Other physical models may look and act more like or less like the thing they represent than the model in **Figure 1** does. Scientists often use the model that is simplest to use but that still serves their purpose.

READING WARM-UP

Objectives

● Give examples of three types of models.

● Identify the benefits and limitations of models.

● Compare the ways that scientists use hypotheses, theories, and laws.

Terms to Learn

model
theory
law

READING STRATEGY

Reading Organizer As you read this section, create an outline of the section. Use the headings from the section in your outline.

model a pattern, plan, representation, or description designed to show the structure or workings of an object, system, or concept

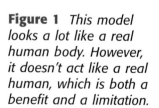

Figure 1 *This model looks a lot like a real human body. However, it doesn't act like a real human, which is both a benefit and a limitation.*

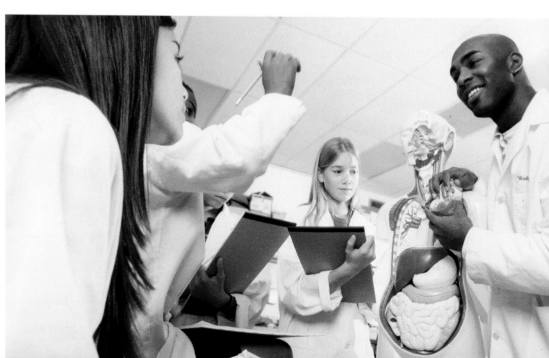

Figure 2 **Mathematical Model: A Punnett Square**

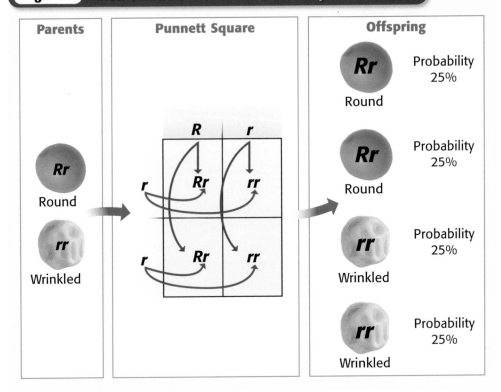

Parents	Punnett Square	Offspring

Mathematical Models

A mathematical model may be made up of numbers, equations, or other forms of data. Some mathematical models are simple and can be used easily. The Punnett square shown in **Figure 2** is a model of how traits may be passed from parents to offspring. Using this model, scientists can predict how often certain traits will appear in the offspring of certain parents.

Computers are very useful for creating and manipulating mathematical models. They make fewer mistakes and can keep track of more variables than a human can. But a computer model can be incorrect in many ways. The more complex a model is, the more carefully scientists must build the model.

Reading Check What type of model is a Punnett square? (*See the Appendix for answers to Reading Checks.*)

Conceptual Models

The third type of model is the conceptual model. Some conceptual models represent systems of ideas. Others compare unfamiliar things with familiar things. These comparisons help explain unfamiliar ideas. The idea that life originated from chemicals is a conceptual model. Scientists also use conceptual models to classify behaviors of animals. Scientists can then predict how an animal might respond to a certain action based on the behaviors that have already been observed.

CONNECTION TO
Social Studies

Gregor Mendel About 150 years ago, Gregor Mendel studied the passing of traits in pea plants. After studying biology at a university, Mendel entered a monastery. His work with peas in the monastery garden started the field of life science called *genetics*.

Use the library or the Internet to research Gregor Mendel. Also, research the time and place where he lived. Discuss the following questions with your classmates: What made Mendel unique in his time? In what ways was Mendel a great life scientist?

ACTIVITY

Figure 3 *This computer-generated model doesn't just look like a dinosaur. This model includes the movement of bones and muscles.*

theory an explanation that ties together many hypotheses and observations

law a summary of many experimental results and observations; a law tells how things work

Benefits of Models

Models are often used to represent things that are very small or very large. Models may also represent things that are very complicated or things that no longer exist. For example, **Figure 3** is a model of a dinosaur. Such computer models have been used for many things, including to make movies about prehistoric life on Earth. Models are used, of course, because filming a real dinosaur in action is impossible. But in building models, scientists may discover things they hadn't thought of before.

A model can be a kind of hypothesis and can be tested. To build a model of an organism, scientists must gather information learned from fossils and other observations. Then, they can test whether their model fits with their ideas about how an organism might have moved or what it might have eaten.

Building Scientific Knowledge

Sometimes, scientists may draw different conclusions from the same data. Other times, new results show that old conclusions are wrong. Sometimes, more information is needed. Life scientists are always asking new questions or looking at old questions from a new angle. As they find new answers, scientific knowledge continues to grow and change.

Scientific Theories

For every hypothesis, more than one prediction can be made. Each time another prediction is proven true, the hypothesis gains more support. Over time, scientists try to tie together all they have learned. An explanation that ties together many related facts, observations, and tested hypotheses is called a **theory.** Theories are conceptual models that help to organize scientific thinking. Theories are used to explain observations and also to predict what might happen in the future.

✓ Reading Check How do scientists use theories?

Scientific Laws

The one kind of scientific idea that rarely changes is called a *scientific law*. In science, a **law** is a summary of many experimental results and observations. Unlike traffic laws, scientific laws are not based on what people may want to happen. Instead, scientific laws are statements of what *will* happen in a specific situation. And unlike theories, scientific laws tell you only what happens, not why it happens.

Combining Scientific Ideas

Scientific laws are at work around you every day. For example, the law of gravity is at work when we see a leaf fall to the ground. The law of gravity tells us that objects always fall toward the center of the Earth. Many laws of chemistry are at work inside your cells. However, living organisms are very complex. So, there are very few laws within life science. But some theories are very important in life science and are widely accepted. An example is the theory that all living things are made up of cells.

Scientific Change

History shows that new scientific ideas take time to develop into theories or to become accepted as facts or laws. Scientists should be open to new ideas, but they should always test those ideas with scientific methods. And if new evidence contradicts an accepted idea, scientists must be willing to re-examine the evidence and re-evaluate their reasoning. The process of building scientific knowledge never ends.

CONNECTION TO Physics

The Laws of Physics Part of understanding a scientific law is knowing the conditions under which it is true. Many of the laws of physics deal with a simple set of conditions. For example, Newton's laws of motion are used to predict how objects, such as planets, will move through space. The same laws apply on Earth, but predicting the motion of objects on Earth is more complex. Look up Newton's laws, and then brainstorm ways in which the conditions in space differ from the conditions on Earth.

ACTIVITY

SECTION Review

Summary

- A model is a representation of an object or system. Models often use familiar things to represent unfamiliar things. Three main types of models are physical, mathematical, and conceptual. Models have limitations but are useful and can be changed based on new evidence.

- Scientific knowledge is built as scientists form and revise scientific hypotheses, models, theories, and laws.

Using Key Terms

In each of the following sentences, replace the incorrect term with the correct term from the word bank.

theory   law

1. A conclusion is an explanation that matches many hypotheses but may still change.

2. A model tells you exactly what to expect in certain situations.

Understanding Key Ideas

3. A limitation of models is that
   a. they are large enough to see.
   b. they do not act exactly like the things that they model.
   c. they are smaller than the things that they model.
   d. they model unfamiliar things.

4. What are three types of models? Give an example of each type.

5. Compare how scientists use theories with how they use laws.

Math Skills

6. If Jerry is 2.1 m tall, how tall is a scale model of Jerry that is 10% of his size?

Critical Thinking

7. **Applying Concepts** You are making a three-dimensional model of an extinct plant. Describe some of the potential uses for your model. What are some limitations of your model?

SCILINKS

NSTA
Developed and maintained by the
National Science Teachers Association

For a variety of links related to this chapter, go to www.scilinks.org

Topic: Using Models
SciLinks code: HSM1588

Tools, Measurement, and Safety

Would you use a hammer to tighten a bolt on a bicycle? You probably wouldn't. To be successful in many tasks, you need the correct tools.

Life scientists use various tools to help them in their work. These tools are used to make observations and to gather, store, and analyze information. Choosing and using tools properly are important parts of scientific work.

Computers and Technology

The application of science for practical purposes is called **technology.** By using technology, life scientists are able to find information and solve problems in new ways. New technology allows scientists to get information that wasn't available previously.

Since the first electronic computer was built in 1946, improvements in technology have made computers more powerful and easier to use. Computers can be used to create graphs, solve complex equations, and analyze data. Computers also help scientists share data and ideas with each other and publish reports about their research.

Tools for Seeing

It's difficult to make accurate observations of things that cannot be seen. When the first microscopes were invented, scientists were able to see into a whole new world. Today, the workings of tiny cells and organisms are well understood. New tools and technologies allow us to see inside organisms in new ways. For example, the images shown in **Figure 1** were created by sending electromagnetic waves through human bodies.

READING WARM-UP

Objectives

● Give three examples of how life scientists use computers and technology.

● Describe three tools life scientists use to observe organisms.

● Explain the importance of the International System of Units, and give four examples of SI units.

Terms to Learn

technology
compound light microscope
electron microscope
area
volume
mass
temperature

READING STRATEGY

Reading Organizer As you read this section, make a concept map by using the terms above.

technology the application of science for practical purposes; the use of tools, machines, materials, and processes to meet human needs

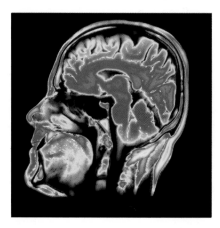
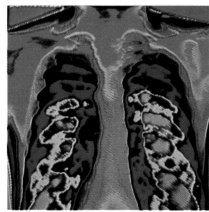

Figure 1 *The image on the left is a computerized axial tomography scan (CAT scan). The image on the right was made with magnetic resonance imagery (MRI).*

Figure 2 Types of Microscopes

Compound Light Microscope
Light passes through the specimen and produces a flat image.

Transmission Electron Microscope Electrons pass through the specimen and produce a flat image.

Scanning Electron Microscope Electrons bounce off the surface of the specimen and produce a three-dimensional (3-D) image.

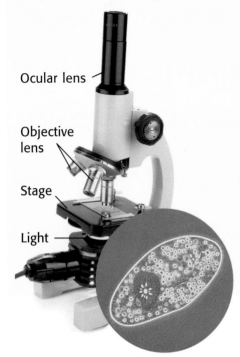

Ocular lens

Objective lens

Stage

Light

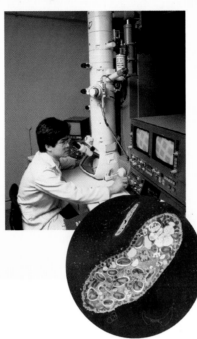

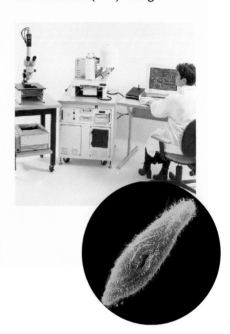

Compound Light Microscope

The compound light microscope is a common tool in a life science laboratory. A **compound light microscope** is an instrument that magnifies small objects so that they can be seen easily. It has three main parts—a tube with two or more lenses, a stage, and a light. Items viewed through a compound microscope may be colored with special dyes to make them more visible. Items are placed on the stage so that the light passes through them. The lenses at each end of the tube magnify the image.

Electron Microscopes

Not all microscopes use light. In **electron microscopes,** tiny particles called *electrons* are used to produce magnified images. The images produced are clearer and more detailed than those made by light microscopes. However, living things cannot be viewed with electron microscopes because the preparation process kills them. There are two kinds of electron microscopes used in life science—the transmission electron microscope (TEM) and the scanning electron microscope (SEM). **Figure 2** shows each kind of microscope, describes the specialized purpose of each, and shows an example of the images each can produce.

compound light microscope
an instrument that magnifies small objects so that they can be seen easily by using two or more lenses

electron microscope a microscope that focuses a beam of electrons to magnify objects

✓ *Reading Check* How are SEMs different from TEMs? (*See the Appendix for answers to Reading Checks.*)

Table 1	Common SI Units and Conversions	
Length	**meter (m)**	
	kilometer (km)	1 km = 1,000 m
	decimeter (dm)	1 dm = 0.1 m
	centimeter (cm)	1 cm = 0.01 m
	millimeter (mm)	1 mm = 0.001 m
	micrometer (μm)	1 μm = 0.000001 m
	nanometer (nm)	1 nm = 0.000000001 m
Volume	**cubic meter (m³)**	
	cubic centimeter (cm³)	1 cm³ = 0.000001 m³
	liter (L)	1 L = 1 dm³ = 0.001 m³
	milliliter (mL)	1 mL = 0.001 L = 1 cm³
Mass	**kilogram (kg)**	
	gram (g)	1 g = 0.001 kg
	milligram (mg)	1 mg = 0.000001 kg
Temperature	**kelvin (K)**	
	Celsius (°C)	0°C = 273 K
		100°C = 373 K

SCHOOL to HOME

How You Measure Matters

Measure the length and width of a desk or table, but do not use a ruler. Pick a common object to use as your unit of measurement. It could be a pencil, your hand, or anything else. Use that unit to determine the area of the desk or table.

To calculate the area of a rectangle, first measure the length and width. Then, use the following equation:

area = length × width

Ask your parent or sibling to do this activity on their own. When they are finished, compare your area calculation with theirs.

Measurement

The ability to make reliable measurements is an important skill in science. But different standards of measurement have developed throughout the world. Ancient measurement units were based on parts of the body, such as the foot, or on objects, such as grains of wheat. Such systems were not very reliable. Even as better standards were developed, they varied from country to country.

The International System of Units

In the late 1700s, the French Academy of Sciences began to form a global measurement system now known as the *International System of Units* (also called *SI*, or *Système International d'Unités*). Today, most scientists and almost all countries use this system. One advantage of using SI measurements is that it helps scientists share and compare their observations and results.

Another advantage of SI units is that almost all units are based on the number 10, which makes conversions from one unit to another easier. **Table 1** contains commonly used SI units for length, volume, mass, and temperature. Notice how the prefix of each SI unit relates to a base unit.

Length

How long is an ant? A life scientist would probably use millimeters (mm) to describe an ant's length. If you divide 1 m into 1,000 parts, each part equals 1 mm. So, 1 mm is one-thousandth of a meter. Although millimeters seem small, some organisms and structures are so tiny that even smaller units—micrometers (μm) or nanometers (nm)—must be used.

Area

How much paper would you need to cover your desktop? To answer this question, you must find the area of the desk. **Area** is a measure of how much surface an object has. Area can be calculated from measurements such as length and width. Area is stated in square units, such as square meters (m^2), square centimeters (cm^2), and square kilometers (km^2).

area a measure of the size of a surface or a region

volume a measure of the size of a body or region in three-dimensional space

✔️ *Reading Check* **What kinds of units describe area?**

Volume

How many books will fit into a backpack? The answer depends on the volume of the backpack and the volume of each book. **Volume** is a measure of the size of something in three-dimensional space.

The volume of a liquid is most often described in liters (L). Liters are based on the meter. A cubic meter ($1 m^3$) is equal to 1,000 L. So 1,000 L will fit into a box measuring 1 m on each side. A milliliter (mL) will fit into a box that is 1 cm on each side. So, $1 mL = 1 cm^3$. Graduated cylinders are used to measure the volume of liquids, as shown in **Figure 3.**

The volume of a solid object is given in cubic units, such as cubic meters (m^3), cubic centimeters (cm^3), or cubic millimeters (mm^3). To find the volume of a box-shaped object, multiply the object's length by its width and height. As **Figure 3** shows, the volume of an irregularly shaped object is found by measuring the volume of liquid that the object displaces.

Figure 3 *A rock added to a graduated cylinder raised the level of water from 70 mL to 80 mL of water. Because the rock displaced 10 mL of water and because 1 mL = 1 cm³, the volume of the rock is 10 cm³.*

70 mL

80 mL

Measure Up!

1. For each of the following tasks, find a different item to measure. With permission from your teacher or parent, you may choose items within your classroom, school, or home.

   a. Measure length with a **meterstick.**

   b. Measure length with a **metric ruler.**

   c. Measure and calculate area in square meters.

   d. Measure volume with a **graduated cylinder.**

   e. Measure and calculate volume in cubic meters.

   f. Measure mass with a **balance.**

   g. Measure temperature with a **thermometer.**

2. Make a **poster** to present your measurements. Include drawings showing how you measured each item and tips stating how to use the measurement tools properly.

mass a measure of the amount of matter in an object

temperature a measure of how hot (or cold) something is

Mass

How much matter is in an apple? **Mass** is a measure of the amount of matter in an object. The kilogram (kg) is the basic unit for mass. The mass of a very large object is described in kilograms or metric tons. A metric ton equals 1,000 kg. The mass of a small object may be described in grams (g). A kilogram equals 1,000 g; therefore, a gram is one-thousandth of a kilogram. A medium-sized apple has a mass of about 100 g. Mass can be measured by using a balance.

Temperature

How much should food be heated to kill any bacteria in the food? To answer this question, a life scientist would measure the temperature at which bacteria die. **Temperature** is a measure of how hot or cold something is. Temperature is actually an indication of the amount of energy within matter. You are probably used to describing temperature in degrees Fahrenheit (°F). Scientists commonly use degrees Celsius (°C), although the kelvin (K) is the official SI base unit for temperature. You will use degrees Celsius in this book. The thermometer in **Figure 4** shows how two of these scales compare.

Reading Check What does temperature indicate about matter?

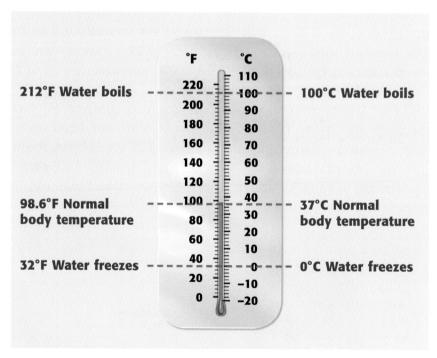

Figure 4 *Water freezes at 0°C and boils at 100°C. Your normal body temperature is 37°C, which is equal to 98.6°F.*

Safety Rules!

Life science is exciting and fun, but it can also be dangerous. So, don't take any chances! Always follow your teacher's instructions, and don't take shortcuts—even when you think there is no danger in doing so. Before starting an experiment, get your teacher's permission, and read the lab procedures carefully. Pay particular attention to safety information and caution statements. **Figure 5** shows the safety symbols used in this book. Get to know these symbols and their meanings by reading the safety information in the front of this book. **This is important!** If you are still unsure about what a safety symbol means, ask your teacher.

Figure 5 **Safety Symbols**

 Eye protection

 Clothing protection

 Hand safety

 Heating safety

 Electric safety

 Sharp object

 Chemical safety

 Animal safety

Plant safety

SECTION Review

Summary

- Life scientists use computers to collect, store, organize, analyze, and share data.
- Life scientists commonly use light microscopes and electron microscopes to make observations of things that are too small to be seen without help. Electromagnetic waves are also used in other ways to create images.
- The International System of Units (SI) is a simple and reliable system of measurement that is used by most scientists.

Using Key Terms

Complete each of the following sentences by choosing the correct term from the word bank.

mass	area
volume	temperature

1. The measure of the surface of an object is called ___.

2. Life scientists use kilograms when measuring an object's ___.

3. The ___ of a liquid is usually described in liters.

Understanding Key Ideas

4. SI units are
   a. always based on standardized measurements of body parts.
   b. almost always based on the number 10.
   c. used only to measure length.
   d. used only in France.

5. How is temperature related to energy?

6. If you were going to measure the mass of a fly, which SI unit would be most appropriate?

Math Skills

7. Convert 3.0 L into cubic centimeters.

8. Calculate the volume of a textbook that is 28.5 cm long, 22 cm wide, and 3.5 cm thick.

Critical Thinking

9. **Making Inferences** The mite shown below is about 500 µm long in real life. What tool was probably used to produce this image? How can you tell?

10. **Applying Concepts** Give an example of what could happen if you do not follow safety rules.

SCiLINKS® NSTA
Developed and maintained by the National Science Teachers Association

For a variety of links related to this chapter, go to www.scilinks.org

Topic: Tools of Life Science; SI Units
SciLinks code: HSM1535; HSM1390

Skills Practice Lab

OBJECTIVES

Apply scientific methods to predict, measure, and observe the mixing of two unknown liquids.

MATERIALS

- beakers, 100 mL (2)
- Celsius thermometer
- glass-labeling marker
- graduated cylinders, 50 mL (3)
- liquid A, 75 mL
- liquid B, 75 mL
- protective gloves

SAFETY

Does It All Add Up?

Your math teacher won't tell you this, but did you know that sometimes 2 + 2 does not appear to equal 4?! In this experiment, you will use scientific methods to predict, measure, and observe the mixing of two unknown liquids. You will learn that a scientist does not set out to prove a hypothesis but to test it and that sometimes the results just don't seem to add up!

Make Observations

1 Put on your safety goggles, gloves, and lab apron. Examine the beakers of liquids A and B provided by your teacher. Write down as many observations as you can about each liquid. **Caution:** Do not taste, touch, or smell the liquids.

2 Pour exactly 25 mL of liquid A from the beaker into each of two 50 mL graduated cylinders. Combine these samples in one of the graduated cylinders. Record the final volume. Pour the liquid back into the beaker of liquid A. Rinse the graduated cylinders. Repeat this step for liquid B.

Form a Hypothesis

3 Based on your observations and on prior experience, formulate a testable hypothesis that states what you expect the volume to be when you combine 25 mL of liquid A with 25 mL of liquid B.

4 Make a prediction based on your hypothesis. Use an if-then format. Explain why you made your prediction.

Data Table				
	Contents of cylinder A	Contents of cylinder B	Mixing results: predictions	Mixing results: observations
Volume				
Appearance		DO NOT WRITE IN BOOK		
Temperature				

Test the Hypothesis

5 Make a data table like the one above.

6 Mark one graduated cylinder "A." Carefully pour exactly 25 mL of liquid A into this cylinder. In your data table, record its volume, appearance, and temperature.

7 Mark another graduated cylinder "B." Carefully pour exactly 25 mL of liquid B into this cylinder. Record its volume, appearance, and temperature in your data table.

8 Mark the empty third cylinder "A + B."

9 In the "Mixing results: predictions" column in your table, record the prediction you made earlier. Each classmate may have made a different prediction.

10 Carefully pour the contents of both cylinders into the third graduated cylinder.

11 Observe and record the total volume, appearance, and temperature in the "Mixing results: observations" column of your table.

Analyze the Results

1 **Analyzing Data** Discuss your predictions as a class. How many different predictions were there? Which predictions were supported by testing? Did any measurements surprise you?

Draw Conclusions

2 **Drawing Conclusions** Was your hypothesis supported or disproven? Either way, explain your thinking. Describe everything that you think you learned from this experiment.

3 **Analyzing Methods** Explain the value of incorrect predictions.

Chapter Review

USING KEY TERMS

1 Use the following terms in the same sentence: *life science* and *scientific methods*.

2 Use the following terms in the same sentence: *controlled experiment* and *variable*.

For each pair of terms, explain how the meanings of the terms differ.

3 *theory* and *hypothesis*

4 *compound light microscope* and *electron microscope*

5 *area* and *volume*

UNDERSTANDING KEY IDEAS

Multiple Choice

6 The steps of scientific methods
  a. must all be used in every scientific investigation.
  b. must always be used in the same order.
  c. often start with a question.
  d. always result in the development of a theory.

7 In a controlled experiment,
  a. a control group is compared with one or more experimental groups.
  b. there are at least two variables.
  c. all factors should be different.
  d. a variable is not needed.

8 Which of the following tools is best for measuring 100 mL of water?
  a. 10 mL graduated cylinder
  b. 150 mL graduated cylinder
  c. 250 mL beaker
  d. 500 mL beaker

9 Which of the following is NOT an SI unit?
  a. meter
  b. foot
  c. liter
  d. kilogram

10 A pencil is 14 cm long. How many millimeters long is it?
  a. 1.4 mm
  b. 140 mm
  c. 1,400 mm
  d. 1,400,000 mm

11 The directions for a lab include the safety icons shown below. These icons mean that

  a. you should be careful.
  b. you are going into the laboratory.
  c. you should wash your hands first.
  d. you should wear safety goggles, a lab apron, and gloves during the lab.

Short Answer

12 List three ways that science is beneficial to living things.

13 Why do hypotheses need to be testable?

14 Give an example of how a life scientist might use computers and technology.

15 List three types of models, and give an example of each.

16 What are some advantages and limitations of models?

17 Which SI units can be used to describe the volume of an object? Which SI units can be used to describe the mass of an object?

18 In a controlled experiment, why should there be several individuals in the control group and in each of the experimental groups?

CRITICAL THINKING

19 **Concept Mapping** Use the following terms to create a concept map: *observations, predictions, questions, controlled experiments, variable,* and *hypothesis.*

20 **Making Inferences** Investigations often begin with observation. What limits are there to the observations that scientists can make?

21 **Forming Hypotheses** A scientist who studies mice observes that on the day the mice are fed vitamins with their meals, they perform better in mazes. What hypothesis would you form to explain this phenomenon? Write a testable prediction based on your hypothesis.

INTERPRETING GRAPHICS

The pictures below show how an egg can be measured by using a beaker and water. Use the pictures to answer the questions that follow.

Before: 125 mL   After: 200 mL

22 What kind of measurement is being taken?

a. area

b. length

c. mass

d. volume

23 Which of the following is an accurate measurement of the egg in the picture?

a. 75 cm^3

b. 125 cm^3

c. 125 mL

d. 200 mL

24 Make a double line graph from the data in the following table.

Number of Frogs		
Date	Normal	Deformed
1995	25	0
1996	21	0
1997	19	1
1998	20	2
1999	17	3
2000	20	5

Standardized Test Preparation

Read each of the passages below. Then, answer the questions that follow the passage.

Passage 1 Zoology is the study of animals. Zoology dates back more than 2,300 years, to ancient Greece. There, the philosopher Aristotle observed and theorized about animal behavior. About 200 years later, Galen, a Greek physician, began dissecting and experimenting with animals. However, there were few advances in zoology until the 1700s and 1800s. During this period, the Swedish <u>naturalist</u> Carolus Linnaeus developed a classification system for plants and animals, and British naturalist Charles Darwin published his theory of evolution by natural selection.

1. According to the passage, when did major advances in Zoology begin?
   A About 2,300 years ago
   B About 2,100 years ago
   C During the 1700s and 1800s
   D Only during recent history

2. Which of the following is a possible meaning of the word *naturalist,* as used in the passage?
   F a scientist who studies plants and animals
   G a scientist who studies animals
   H a scientist who studies theory
   I a scientist who studies animal behavior

3. Which of the following is the **best** title for this passage?
   A Greek Zoology
   B Modern Zoology
   C The Origins of Zoology
   D Zoology in the 1700s and 1800s

Passage 2 When looking for answers to a problem, scientists build on existing knowledge. For example, scientists have wondered if there is some relationship between Earth's core and Earth's magnetic field. To form a hypothesis, scientists started with what they knew: Earth has a dense, solid inner core and a molten outer core. Scientists then created a computer <u>model</u> to simulate how Earth's magnetic field might be generated.

They tried different things with their model until the model produced a magnetic field that matched that of the real Earth. The model predicted that Earth's inner core spins in the same direction as the rest of the Earth, but the inner core spins slightly faster than Earth's surface. If the hypothesis is correct, it might explain how Earth's magnetic field is produced. Although scientists cannot reach the Earth's core to examine it directly, they can test whether other observations match what is predicted by their hypothesis.

1. What does the word *model* refer to in this passage?
   A a giant plastic globe
   B a representation of the Earth created on a computer
   C a computer terminal
   D a technology used to drill into the Earth's core

2. Which of the following is the **best** summary of the passage?
   F Scientists can use models to help them answer difficult and complex questions.
   G Scientists have discovered the source of Earth's magnetic field.
   H The spinning of Earth's molten inner core causes Earth's magnetic field.
   I Scientists make a model of a problem and then ask questions about the problem.

The table below shows the plans for an experiment in which bees will be observed visiting flowers. Use the table to answer the questions that follow.

Bee Experiment				
Group	Type of bee	Time of day	Type of plant	Flower color
#1	Honey-bee	9:00 A.M.–10:00 A.M.	Portland rose	red
#2	Honey-bee	9:00 A.M.–10:00 A.M.	Portland rose	yellow
#3	Honey-bee	9:00 A.M.–10:00 A.M.	Portland rose	white
#4	Honey-bee	9:00 A.M.–10:00 A.M.	Portland rose	pink

1. Which factor is the variable in this experiment?
   A the type of bee
   B the time of day
   C the type of plant
   D the color of the flowers

2. Which of the following hypotheses could be tested by this experiment?
   F Honeybees prefer to visit rose plants.
   G Honeybees prefer to visit red flowers.
   H Honeybees prefer to visit flowers in the morning.
   I Honey bees prefer to visit Portland rose flowers between 9 and 10 A.M.

3. Which of the following is the **best** reason why the Portland rose plant is included in all of the groups to be studied?
   A The type of plant is a control factor; any type of flowering plant could be used as long as all plants were of the same type.
   B The experiment will test whether bees prefer the Portland rose over other flowers.
   C An experiment should always have more than one variable.
   D The Portland rose is a very common plant.

Read each question below, and choose the best answer.

1. A survey of students was conducted to find out how many people were in each student's family. The replies from five students were as follows: 3, 3, 4, 4, and 6. What was the average family size?
   A 3
   B 3.5
   C 4
   D 5

2. In the survey above, if one more student were surveyed, which reply would make the average lower?
   F 3
   G 4
   H 5
   I 6

3. If an object that is 5 μm long were magnified by 1,000, how long would that object then appear?
   A 5 μm
   B 5 mm
   C 1,000 μm
   D 5,000 mm

4. How many meters are in 50 km?
   F 50 m
   G 500 m
   H 5,000 m
   I 50,000 m

5. What is the area of a square whose sides measure 4 m each?
   A 16 m
   B 16 m^2
   C 32 m
   D 32 m^2

Standardized Test Preparation

Science in Action

HOLT ANTHOLOGY OF
Science Fiction

HOLT, RINEHART AND WINSTON

Scientific Debate

Should We Stop All Forest Fires?

Since 1972, the policy of the National Park Service has been to manage the national parks as naturally as possible. Because fire is a natural event in forests, this policy includes allowing most fires caused by lightning to burn. The only lightning-caused fires that are put out are those that threaten lives, property, uniquely scenic areas, or endangered species. All human-caused fires are put out. However, this policy has caused some controversy. Some people want this policy followed in all public forests and even grasslands. Others think that all fires should be put out.

Science Fiction

"The Homesick Chicken" by Edward D. Hoch

Why did the chicken cross the road? You think you know the answer to this old riddle, don't you? But "The Homesick Chicken," by Edward D. Hoch, may surprise you. That old chicken may not be exactly what it seems.

You see, one of the chickens at the high-tech Tangaway Research Farms has escaped. Then, it was found in a vacant lot across the highway from Tangaway, pecking away contentedly. Why did it bother to escape? Barnabus Rex, a specialist in solving scientific riddles, is called in to work on this mystery. As he investigates, he finds clues and forms a hypothesis. Read the story, and see if you can explain the mystery before Mr. Rex does.

Social Studies ACTIVITY

WRITING SKILL Research a location where there is a debate about controlling forest fires. You might look into national forests or parks. Write a newspaper article about the issue. Be sure to present all sides of the debate.

Language Arts ACTIVITY

WRITING SKILL Write your own short story about a chicken crossing a road for a mysterious reason. Give clues (evidence) to the reader about the mysterious reason but do not reveal the truth until the end of the story. Be sure the story makes sense scientifically.

People in Science

Yvonne Cagle

Flight Surgeon and Astronaut Most doctors practice medicine with both feet on the ground. But Dr. Yvonne Cagle found a way to fly with her medical career. Cagle became a flight surgeon for the United States Air Force and an astronaut for the National Aeronautics and Space Administration (NASA).

Cagle's interest in both medicine and space flight began early. As a little girl, Cagle spent hours staring at X rays in her father's medical library. Those images sparked an early interest in science. Cagle also remembers watching Neil Armstrong walk on the moon when she was five years old. As she tried to imagine the view of Earth from space, Cagle decided she wanted to see it for herself.

Becoming an Air Force flight surgeon was a good first step toward becoming an astronaut. As a flight surgeon, Cagle learned about the special medical challenges humans face when they are launched high above the Earth. Being a flight surgeon had the extra benefits of working with some of the best pilots and getting to fly in the latest jets.

It wasn't long before Cagle worked as an occupational physician for NASA at the Johnson Space Center. Two years later, she was chosen to begin astronaut training. Cagle is looking forward to her first flight into space. Her first mission will likely take her to the *International Space Station*, where she can monitor astronaut health and perform scientific experiments.

Math ACTIVITY

In space flight, astronauts experience changes in gravity that affect their bodies in several ways. Because of gravity, a person who has a mass of 50 kg weighs 110 pounds on Earth. But on the moon, the same person weighs about 17% of his or her weight on Earth. How much does the same person weigh on the moon?

To learn more about these Science in Action topics, visit go.hrw.com and type in the keyword HL5LIVF.

Current Science

Check out Current Science® articles related to this chapter by visiting go.hrw.com. Just type in the keyword HL5CS01.

go.hrw.com

2

The World of Earth Science

About the PHOTO

What is that man doing? Ricardo Alonso, a geologist in Argentina, is measuring the footprints left by a dinosaur millions of years ago. Taking measurements is just one way that scientists collect data to answer questions and test hypotheses.

PRE-READING ACTIVITY

FOLDNOTES **Key-Term Fold** Before you read the chapter, create the FoldNote entitled "Key-Term Fold" described in the **Study Skills** section of the Appendix. Write a key term from the chapter on each tab of the key-term fold. Under each tab, write the definition of the key term.

START-UP ACTIVITY

Mission Impossible?

In this activity, you will do some creative thinking to solve what might seem like an impossible problem.

Procedure

1. Examine an **index card.** Your mission is to fit yourself through the card. You can only tear and fold the card. You cannot use tape, glue, or anything else to hold the card together.

2. Brainstorm with a partner ways to complete your mission. Then, record your plan.

3. Test your plan. Did it work? If necessary, get **another index card** and try again. Record your new plan and the results.

4. Share your plans and results with your classmates.

Analysis

1. Why was it helpful to come up with a plan in advance?

2. How did testing your plan help you complete your mission?

3. How did sharing your ideas with your classmates help you complete your mission? What did your classmates do differently?

Branches of Earth Science

Planet Earth! How can anyone study something as large and complicated as our planet?

One way is to divide the study of the Earth into smaller areas of study. In this section, you will learn about some of the most common areas of study. You will also learn about some of the people that work within these areas.

READING WARM-UP

Objectives

● Describe the four major branches of Earth science.

● Identify four examples of Earth science that are linked to other areas of science.

Terms to Learn

geology   meteorology
oceanography   astronomy

READING STRATEGY

Discussion Read this section silently. Write down questions that you have about this section. Discuss your questions in a small group.

Geology—Science That Rocks

The study of the origin, history, and structure of the Earth and the processes that shape the Earth is called **geology.** Everything that has to do with the solid Earth is part of geology.

Most geologists specialize in a particular aspect of the Earth. For example, a *volcanologist* is a geologist who studies volcanoes. Are earthquakes more to your liking? Then, you could be a *seismologist,* a geologist who studies earthquakes. How about digging up dinosaurs? You could be a *paleontologist,* a geologist who studies fossils. These are only a few of the careers you could have as a geologist.

Some geologists become highly specialized. For example, geologist Robert Fronk, at the Florida Institute of Technology, explores the subsurface of Earth by scuba diving in underwater caves in Florida and the Bahamas. Underwater caves often contain evidence that sea level was once much lower than it is now. The underwater caves shown in **Figure 1** contain *stalagmites* and *stalactites.* These formations develop from minerals in water that drips in air-filled caves. When Fronk sees these kinds of geologic formations in underwater caves, he knows that the caves were once above sea level.

geology the study of the origin, history, and structure of the Earth and the processes that shape the Earth

Figure 1 *Stalagmites grow upward from the floors of caves, and stalactites grow downward from the ceilings of caves.*

Oceanography—Water, Water Everywhere

The scientific study of the sea is called **oceanography.** Special areas of oceanography include physical oceanography, biological oceanography, geological oceanography, and chemical oceanography. Physical oceanographers study physical features of the ocean such as waves and currents to see how they affect weather patterns and aquatic life. Biological oceanographers study the plants and animals that live in the ocean. Geological oceanographers study and explore the ocean floor for clues to the Earth's history. Chemical oceanographers study amounts and distributions of natural and human-made chemicals in the ocean.

oceanography the scientific study of the sea

✓ **Reading Check** Describe four special areas of oceanography. (*See the Appendix for answers to Reading Checks.*)

Exploring the Ocean Floor

Not long ago, people studied the ocean only from the surface. But as technology has advanced, scientists have worked with engineers to build miniature research submarines to go practically anywhere in the oceans.

John Trefry is an oceanographer who studies the ocean floor in a minisub called *Alvin.* Using the *Alvin,* Trefry can travel 2.4 km below the surface of the ocean. At this depth, Trefry can explore an interesting new world. One of the most exciting sights Trefry has seen is a black smoker. As shown in **Figure 2,** *black smokers* are rock chimneys on the ocean floor that spew black clouds of minerals. Black smokers are a kind of *hydrothermal vent,* which is a crack in the ocean floor that releases very hot water from beneath the Earth's surface. The minerals and hot water from these vents support a beautiful and exotic biological community. The aquatic life includes blood-red tube worms that are 3.5 m long, clams that are 30 cm in diameter, and blind white crabs.

How Hot Is 300°C?

1. Use a **thermometer** to measure the air temperature in the room in degrees Celsius. Record your reading.

2. Hold the thermometer near a **heat source** in the room, such as a light bulb or a heating vent. Be careful not to burn yourself. Record your reading.

3. How do the temperatures you recorded compare with the 300°C temperature of the water from a black smoker?

Figure 2 *Black smokers, such as this one seen through the window of* Alvin, *can reach temperatures up to 300°C!*

Figure 3 *This image, made from several satellite photos, traces Hurricane Andrew's path at three locations from the Atlantic Ocean* (right) *to the Gulf of Mexico* (left).

Meteorology—It's a Gas!

The study of the Earth's atmosphere, especially in relation to weather and climate, is called **meteorology.** When you ask, "Is it going to rain today?" you are asking a meteorological question. One of the most common careers in meteorology is weather forecasting. Sometimes, knowing what the weather will be like makes our lives more comfortable. Sometimes, our lives depend on these forecasts.

meteorology the scientific study of the Earth's atmosphere, especially in relation to weather and climate

Hurricanes

In 1928, a major hurricane hit Florida and killed 1,836 people. In contrast, a hurricane of similar strength—Hurricane Andrew, shown in **Figure 3**—hit Florida in 1992 and killed 48 people. Why were there far fewer deaths in 1992? Two major reasons were hurricane tracking and weather forecasting.

Meteorologists began tracking Hurricane Andrew on Monday, August 17, 1992. By the following Sunday morning, most people in southern Florida had left the coast. The National Hurricane Center had warned them that Andrew was headed their way. The hurricane caused a lot of damage. However, it killed very few people, thanks to meteorologists' warnings.

Figure 4 *These meteorologists are risking their lives to gather data about tornadoes.*

Tornadoes

An average of 780 tornadoes touch down each year in the United States. What do you think about a meteorologist who chases tornadoes as a career? Howard Bluestein does just that. He predicts where tornadoes are likely to form. He then drives to within a couple of kilometers of the site to gather data, as shown in **Figure 4.** By gathering data this way, scientists such as Bluestein hope to understand tornadoes better. The better scientists understand tornadoes, the better scientists can predict how tornadoes will behave.

Astronomy—Far, Far Away

How do you study things that are beyond Earth? Astronomers can answer this question. **Astronomy** is the study of the universe. Astronomers study stars, asteroids, planets, and everything else in space.

Because most things in space are too far away to study directly, astronomers depend on technology to help them study objects in space. Optical telescopes are one way astronomers study objects in space. Optical telescopes have been used for hundreds of years. Galileo built an optical telescope in 1609. But optical telescopes are not the only kind of telescope astronomers use.

Optical telescopes need light to see objects, such as planets and comets. However, some objects do not give off light or are too far away to be seen with an optical telescope. Instead of detecting the visible light waves, radio telescopes, such as the one in **Figure 5,** detect radio waves. Radio waves are not visible like light waves are, but data from radio waves form patterns. From these patterns, astronomers can make images to learn more about the objects in space.

Star Struck

Astronomers spend much of their time studying stars. Astronomers estimate that there are more than 100 billion billion stars—that is a lot of stars! The most familiar star in the universe is the sun. The sun is the closest star to the Earth. For this reason, astronomers have studied the sun more than other stars.

✓ Reading Check What do astronomers study?

astronomy the study of the universe

Lots of Zeros!

Astronomers estimate that there are more than 100 billion billion stars! One billion written out in numerals looks like this: 1,000,000,000.

How many zeros do you need in order to write 100 billion billion in numerals? To find out, multiply 1 billion by 1 billion, and then multiply your answer by 100. Count the zeros in the final answer.

Now, time how long it takes you to count to 100. How long would it take you to count to 100 a billion billion times?

Figure 5 *Radio telescopes receive radio waves from objects in space.*

Figure 6 *This environmental scientist is measuring chemicals in the water to look for traces of urban or industrial pollution.*

Special Branches of Earth Science

In addition to the main branches of Earth science, there are branches that depend heavily on other areas of science. Earth scientists often find themselves in careers that rely on life science, chemistry, physics, and many other areas of science.

Environmental Science

The study of how humans interact with the environment is called *environmental science.* As shown in **Figure 6,** one task of an environmental scientist is to determine how humans affect the environment. Environmental science relies on geology, life science, chemistry, and physics to help preserve Earth's resources and to teach others how to use them wisely.

Ecology

By studying the relationships between organisms and their surroundings, scientists can better understand the behavior of these organisms. An *ecologist* is a person who studies a community of organisms and their nonliving environment. Ecologists work in many fields, such as wildlife management, agriculture, forestry, and conservation.

Geochemistry

Geochemistry combines the studies of geology and chemistry. *Geochemists,* such as the one in **Figure 7,** specialize in the chemistry of rocks, minerals, and soil. By studying the chemistry of these materials, geochemists can determine the economic value of the materials. Geochemists also can determine what the environment was like when the rocks first formed. Additionally, geochemists study the distribution and effect of chemicals added to the environment by human activity.

Figure 7 *This geochemist is taking rock samples from the field so she can perform chemical analyses of them in a laboratory.*

Geography and Cartography

Physical geographers, who are educated in geology, biology, and physics, study the surface features of Earth. *Cartographers* make maps of those features by using aerial and satellite photos, and computer mapping systems. Have you ever wondered why cities are located where they are? Often, the location of a city is determined by geography. Many cities, such as the one in **Figure 8,** were built near bodies of water because boats were used for transporting people and trade items. Rivers and lakes also provide communities with water for drinking and for raising crops and animals.

✔ Reading Check What do cartographers do?

Figure 8 *The Mississippi River helped St. Louis become the large city it is today.*

SECTION Review

Summary

- The four major branches of Earth science are geology, oceanography, meteorology, and astronomy.

- Other areas of science that are linked to Earth science are environmental science, geochemistry, ecology, geography, and cartography.

- Some careers that are associated with branches of Earth science are volcanologist, seismologist, paleontologist, oceanographer, meteorologist, and astronomer.

Using Key Terms

1. Use each of the following terms in a separate sentence: *geology, oceanography,* and *astronomy.*

Understanding Key Ideas

2. Which of the following Earth scientists would study tornadoes?
   a. a geologist
   b. an oceanographer
   c. a meteorologist
   d. an astronomer

3. On which major branch of Earth science does geochemistry rely?
   a. geology
   b. oceanography
   c. meteorology
   d. astronomy

4. List the major branches of Earth science.

5. In which major branch of Earth science would a scientist study black smokers?

6. List two branches of Earth science that rely heavily on other areas of science. Explain how the branches rely on the other areas of science.

7. List and describe three Earth science careers.

Math Skills

8. Each week, a volcanologist reads 80 pages in a book about volcanoes. In a 4-week period, how many pages will the volcanologist read?

Critical Thinking

9. **Making Inferences** If you were a *hydrogeologist,* what kind of work would you do?

10. **Identifying Relationships** Explain why an ecologist might need to understand geology.

11. **Applying Concepts** Explain how an airline pilot would use Earth science in his or her career.

For a variety of links related to this chapter, go to www.scilinks.org

Topic: Branches of Earth Science
SciLinks code: HSM0191

Scientific Methods in Earth Science

Imagine that you are standing in a thick forest on the bank of a river. Suddenly, you hear a booming noise, and you feel the ground begin to shake.

You notice a creature's head looming over the treetops. The creature's head is so high that its neck must be 20 m long! Then, the entire animal comes into view. You now understand why the ground is shaking. The giant animal is *Seismosaurus hallorum* (SIEZ moh SAWR uhs hah LOHR uhm), the "earth shaker," illustrated in **Figure 1.**

Learning About the Natural World

The description of the *Seismosaurus hallorum* is not based on imagination alone. Scientists have been studying dinosaurs since the 1800s. Scientists gather bits and pieces of information about dinosaurs and their environment. Then, they re-create what dinosaurs might have been like 150 million years ago. But how do scientists put it all together? How do they know if they have discovered a new species? Asking questions like these is the beginning of a process scientists use to learn more about the natural world.

✓ Reading Check How do scientists begin to learn about the natural world? *(See the Appendix for answers to Reading Checks.)*

READING WARM-UP

Objectives

- Explain how scientists begin to learn about the natural world.
- Explain what scientific methods are and how scientists use them.
- Identify the importance of communicating the results of a scientific investigation.
- Describe how scientific investigations often lead to new investigations.

Terms to Learn

scientific methods
hypothesis

READING STRATEGY

Mnemonics As you read this section, create a mnemonic device to help you remember the steps of scientific methods.

Figure 1 Seismosaurus hallorum *is one of the largest dinosaurs known.*

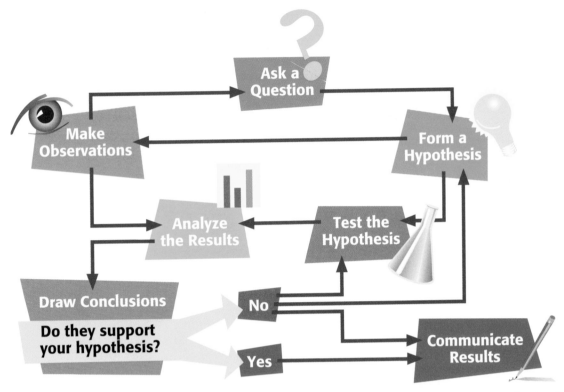

Figure 2 *Steps of scientific methods are illustrated in this flowchart. Notice that there are several ways to follow the paths.*

The flowchart contains the following steps: Ask a Question, Form a Hypothesis, Make Observations, Test the Hypothesis, Analyze the Results, Draw Conclusions — Do they support your hypothesis?, No, Yes, Communicate Results.

What Are Scientific Methods?

When scientists observe the natural world, they often think of a question or problem. But scientists don't just guess answers. Instead, they follow a series of steps called *scientific methods.* **Scientific methods** are a series of steps that scientists use to answer questions and solve problems. The most basic steps are shown in **Figure 2.**

Although scientific methods have several steps, there is not a set procedure. Scientists may use all of the steps or just some of the steps. They may even repeat some of the steps or do them in a different order. The goal of scientific methods is to come up with reliable answers and solutions. Scientists use scientific methods to gain insight into the problems they investigate.

scientific methods a series of steps followed to solve problems

Ask a Question

Asking a question helps focus the purpose of an investigation. For example, David D. Gillette, a scientist who studies fossils, examined some bones found by hikers in New Mexico in 1979. He could tell they were bones of a dinosaur. But he didn't know what kind of dinosaur. Gillette may have asked, "What kind of dinosaur did these bones come from?" Gillette knew that in order to answer this question, he would have to use scientific methods.

Form a Hypothesis

hypothesis an explanation that is based on prior scientific research or observations and that can be tested

When scientists want to investigate a question, they form a hypothesis (hie PAHTH uh sis). A **hypothesis** is a possible explanation or answer to a question that can be tested. Based on his observations and on what he already knew, Gillette said that the bones, shown in **Figure 3,** came from a kind of dinosaur not yet known to scientists. This hypothesis was Gillette's best testable explanation for what kind of dinosaur the bones came from.

Test the Hypothesis

Once a hypothesis is formed, it must be tested. Scientists test hypotheses by gathering data. The data can help scientists tell if the hypotheses are valid or not. To test his hypothesis, Gillette studied the dinosaur bones.

Controlled Experiments

For another activity related to this chapter, go to **go.hrw.com** and type in the keyword **HZ5WESW.**

To test a hypothesis, a scientist may do a controlled experiment. A *controlled experiment* is an experiment that tests only one factor, or *variable,* at a time. All other variables remain constant. By changing only one variable, scientists can see the results of just that one change. If more than one variable is changed, scientists cannot easily determine which variable caused the outcome. For example, let's say you tried to make a gelatin fruit mold, but the gelatin would not harden. The next time you made the gelatin fruit mold, you take out the oranges and pineapples. The gelatin might harden this time, but you won't know whether the pineapples or the oranges caused the gelatin not to harden the first time.

Figure 3 *Gillette and his team had to carefully dig out the bones before taking them to the laboratory for further study.*

Making Observations

Controlled experiments are important for testing hypotheses. Some scientists, however, often depend more on observations than experiments to test their hypotheses. Because scientists cannot always control all variables, some scientists often observe nature and collect large amounts of data. Gillette took hundreds of measurements of the dinosaur bones, as illustrated in **Figure 4.** He compared his measurements with those of bones from known dinosaurs. He also visited museums and talked with other scientists.

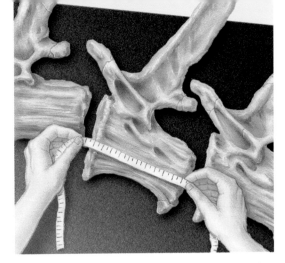

Figure 4 *Gillette observed and measured the dinosaur bones to test his hypothesis.*

Keeping Accurate Records

When testing a hypothesis, a scientist's expectations can affect what he or she actually observes. For this reason, it is important for scientists to keep clear, honest, and accurate records of their experiments and observations. Scientists should present findings supported by scientific data, not by opinions. When possible, scientists will repeat experiments to verify their findings. A hypothesis cannot be examined usefully in a scientific way without enough data. Just one example is never enough to prove something true. However, one example could prove that something is not true.

Analyze the Results

Once scientists finish their tests, they must analyze the results. Scientists often make tables and graphs to organize and summarize their data. When Gillette analyzed his results, he found that the bones of the mystery dinosaur did not match the bones of any known dinosaur. The bones were either too large or too different in shape.

✓ Reading Check Why would scientists create graphs and tables of their data?

CONNECTION TO Oceanography

WRITING SKILL **Making Hypotheses** Scientists exploring the Texas Gulf Coast have discovered American Indian artifacts that are thousands of years old. The odd thing is that the artifacts were buried in the sea floor several meters below sea level. These artifacts had not been moved since they were originally buried. If American Indian artifacts are several meters below sea level, the question to ask is, "Why are they there?" In your **science journal,** form a hypothesis that answers this question. Remember, your hypothesis must be stated in such a way that it can be tested using scientific methods.

Earth Shaker!

One foot is equal to 0.305 m. If a *Seismosaurus hallorum* was 45 m long, how long is the *Seismosaurus hallorum* in inches?

Draw Conclusions

After carefully analyzing the results of their tests, scientists must conclude whether the results supported the hypothesis. Hypotheses are valuable even if they turn out not to be true. If a hypothesis is not supported by the tests, scientists may repeat the investigation to check for errors. Or they may ask new questions and form new hypotheses.

Based on all his studies, Gillette concluded that the bones found in New Mexico were indeed from an unknown dinosaur. This dinosaur, shown in **Figure 5,** was probably 45 m long and weighed between 60 and 100 tons. The creature certainly fit the name Gillette gave it—*Seismosaurus hallorum,* the "earth shaker."

Communicate Results

After finishing an investigation, scientists communicate their results. In this way, scientists share with others what they have learned. Science depends on the sharing of information. Scientists share information by writing reports for scientific journals and giving lectures on their results.

Gillette shared his discovery of *Seismosaurus* at a press conference at the New Mexico Museum of Natural History and Science. He later sent a report that described his investigation to the *Journal of Vertebrate Paleontology.*

When a scientist reveals new evidence, other scientists will evaluate the evidence. They will review the experimental procedure, the data from the experiments, and the reasoning behind explanations. This questioning of evidence and explanations is part of scientific inquiry. Scientists know that their results may be questioned by other scientists. They also understand that data aren't always interpreted the same way by two people. In some cases, another scientist may have published different results on the same topic. In this case, scientists may come to different conclusions. When there is disagreement, scientists will further investigate to find the truth.

Reading Check Why is it important for the scientific community to review new evidence?

Figure 5 *This model of the skeleton of* Seismosaurus hallorum *is based on Gillette's research. The darker-colored bones are those that have been found so far.*

2 m

Case Closed?

Even after results are reviewed and accepted by the scientific community for publication, the investigation of the topic may not be finished. New evidence may become available. The scientist may change the hypothesis based on the new evidence. In other cases, the scientist may have more questions that arise from the original evidence. For example, with the discovery of the *Seismosaurus*, Gillette may have wondered what the *Seismosaurus* ate. What environment did it live in? How did it become extinct? As shown in **Figure 6,** Gillette continues to use scientific methods to answer these questions.

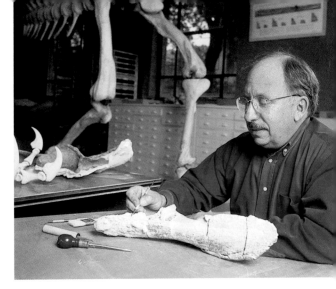

Figure 6 *Gillette continues to study the bones of* Seismosaurus *for new insights into the past.*

SECTION Review

Summary

- Scientists begin to learn about the natural world by asking questions.
- The steps of scientific methods are to ask a question, form a hypothesis, test the hypothesis, analyze the results, draw conclusions, and communicate results.
- Communicating results allows the evidence to be reviewed for accuracy by other scientists.
- Scientific investigations often lead people to ask new questions about the topic.

Using Key Terms

1. Use the following terms in the same sentence: *scientific method* and *hypothesis*.

Understanding Key Ideas

2. Which of the following is NOT part of scientific methods?
   a. ask a question
   b. test the hypothesis
   c. analyze results
   d. close the case

3. Which of the following is the step in scientific methods in which a scientist uses a controlled experiment?
   a. form a hypothesis
   b. test the hypothesis
   c. analyze results
   d. communicate results

4. Explain how scientists use more than imagination to form answers about the natural world.

5. Why do scientists communicate the results of their investigations?

6. For what reason might a scientist change his or her hypothesis after it has already been accepted?

Math Skills

7. If the *Seismosaurus*'s neck is 20 m long and the scientist studying *Seismosaurus* is 2 m long, how many scientists, lined up head to toe, would it take to equal the length of a *Seismosaurus* neck?

Critical Thinking

8. **Applying Concepts** Why might two scientists develop different hypotheses based on the same observations? Explain.

9. **Evaluating Hypotheses** Explain why Gillette's hypothesis—that the bones came from a kind of dinosaur unknown to science— is a testable hypothesis.

SCI LINKS.

NSTA
Developed and maintained by the National Science Teachers Association

For a variety of links related to this chapter, go to www.scilinks.org

Topic: Scientific Methods
SciLinks code: HSM1359

Scientific Models

For your next science project, you will be studying volcanoes. To help you learn more about volcanoes, your teacher suggests using baking soda, vinegar, and clay. How can this help you learn about volcanoes?

Baking soda, vinegar, and clay were the materials used to make the model of the volcano shown in **Figure 1.** By building a model, you can learn more about how volcanoes work.

Types of Scientific Models

A pattern, plan, representation, or description designed to show the structure or workings of an object, system, or concept is a **model.** Models are used to help us understand the natural world. With a model, a scientist can explain or analyze an object, system, or concept in more detail. Models can be used to represent things that are too small to see, such as atoms, or too large to completely see, such as the Earth or the solar system. Models can also be used to explain the past and present and to predict the future. There are three major types of scientific models—physical, mathematical, and conceptual.

Physical Models

Physical models are models that you can touch. Model airplanes, cars, and dolls are all physical models. Physical models often look like the real thing. However, physical models have limitations. For example, a doll is a model of a baby, but a doll does not act like a baby.

READING WARM-UP

Objectives

- Explain how models are used in science.
- Describe the three types of models.
- Identify which types of models are best for certain topics.
- Describe the climate model as an example of a mathematical model.

Terms to Learn

model
theory

READING STRATEGY

Reading Organizer As you read this section, make a table comparing the three types of models.

model a pattern, plan, representation, or description designed to show the structure or workings of an object, system, or concept

Figure 1 *The model volcano looks a little bit like the real volcano, but the model cannot destroy acres of forests with hot lava!*

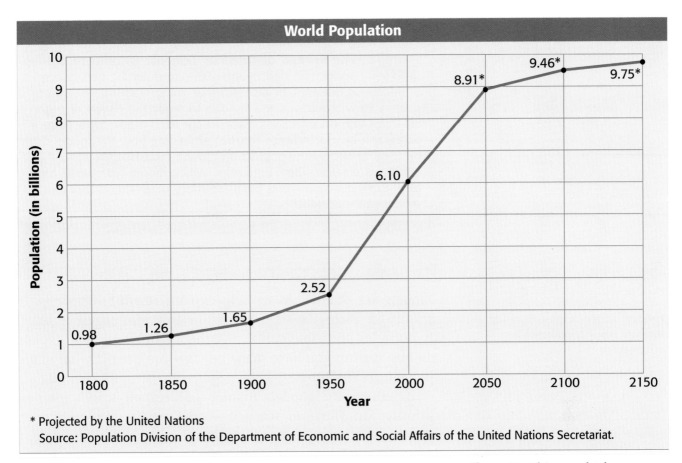

World Population

Population (in billions) — y-axis: 0, 1, 2, 3, 4, 5, 6, 7, 8, 9, 10

Year — x-axis: 1800, 1850, 1900, 1950, 2000, 2050, 2100, 2150

Data points:
- 0.98 (1800)
- 1.26 (1850)
- 1.65 (1900)
- 2.52 (1950)
- 6.10 (2000)
- 8.91* (2050)
- 9.46* (2100)
- 9.75* (2150)

* Projected by the United Nations
Source: Population Division of the Department of Economic and Social Affairs of the United Nations Secretariat.

Figure 2 *This graph shows human population growth predicted by a mathematical model run on a computer.*

Mathematical Models

A *mathematical model* is made up of mathematical equations and data. Some mathematical models are simple. These models allow you to calculate things such as how far a car will go in an hour or how much you would weigh on the moon. Other models are so complex that computers are needed to process them. Look at **Figure 2.** Scientists used a mathematical model to predict population growth in the world. There are many variables that affect population growth. A computer helped process these variables into a model scientists could use.

Conceptual Models

The third type of model is a *conceptual model*. Some conceptual models are systems of ideas. Others are based on making comparisons with familiar things to help illustrate or explain an idea. One example of a conceptual model is the big bang theory. The *big bang theory* is an explanation of the structure of the universe. Conceptual models are composed of many hypotheses. Each hypothesis is supported through scientific methods.

✔ Reading Check What is the big bang theory? (*See the Appendix for answers to Reading Checks.*)

CONNECTION TO
Social Studies

WRITING SKILL **The Spread of Disease** Scientists have found that as population density increases, so does the occurrence of infectious diseases, or diseases that are spread from person to person. How would you use models to study the effects of population growth on infectious disease? For each model type, write a paragraph in your **science journal** explaining how the model might be used to show how population growth increases the spread of infectious disease. Then, determine which model type works the best. Explain why it works best. Are there other things that this model can represent?

Choosing the Right Model

theory an explanation that ties together many hypotheses and observations

Models are often used to help explain scientific theories. In science, a **theory** is an explanation that ties together many hypotheses and observations. A theory not only can explain an observation you have made but also can predict what might happen in the future.

Scientists use models to help guide their search for new information. However, the right model must be chosen in order for the scientist to be able to learn from it. For example, a physical model is useful to understand objects that are too small or too large to see completely. In these cases, a model can help you picture the thing in your mind.

The information that scientists gather by using models can help support a theory or show it to be wrong. Models can be changed or replaced. These changes happen when new observations are made that lead scientists to change their theories. For example, **Figure 3** shows that as scientists' knowledge of the Earth changed, so did the Earth's model.

Figure 3 *Scientists' model of Earth changed as new information was gathered.*

Climate Models

Scientists who study the Earth's atmosphere have developed mathematical climate models to try to imitate Earth's climate. A climate model is like a complicated recipe with thousands of ingredients. One important ingredient is the level of carbon dioxide in the atmosphere. Other ingredients are land and ocean-water temperatures around the globe as well as information about clouds, cloud cover, snow, ice cover, and ocean currents.

You may be wondering how a model can be created with so much data. Because of the development of more powerful computers, scientists are able to process large amounts of data from many different variables, as shown in **Figure 4.** These mathematical models do not make exact predictions about future climates, but they do estimate what might happen. Someday, these models may help scientists prevent serious climate problems, such as global warming or another ice age.

Figure 4 *This meteorologist is using a high-speed supercomputer to do climate modeling.*

✓ Reading Check Why is a climate model complicated?

SECTION Review

Summary

- Models are used to help us understand the natural world.

- There are three types of models: physical models, mathematical models, and conceptual models.

- Scientists must choose the right type of model to learn about a topic.

- A climate model is a mathematical model with so many variables that powerful computers are needed to process the data.

Using Key Terms

1. In your own words, write a definition for each of the following terms: *model* and *theory*.

Understanding Key Ideas

2. Which of the following types of models are systems of ideas?
   a. physical models
   b. mathematical models
   c. conceptual models
   d. climate models

3. Why do scientists use models?

4. Describe the three types of models.

5. Which type of model would you use to study objects that are too small to be seen? Explain.

6. Describe why the climate model is a mathematical model.

Math Skills

7. A model of a bridge is 1 m long and 2.5% of the actual size of the bridge. How long is the actual bridge?

Critical Thinking

8. **Analyzing Ideas** Describe one advantage of physical models.

9. **Applying Concepts** What type of model would you use to study an earthquake? Explain.

SCiLINKS®

NSTA
Developed and maintained by the National Science Teachers Association

For a variety of links related to this chapter, go to www.scilinks.org

Topic: Using Models
SciLinks code: HSM1588

Measurement and Safety

Have you ever used your hand or your foot to measure the length of an object?

At one time, standardized units were based on parts of the body. Long ago, in England, the standard for an inch was three grains of barley. Using such units was not a very accurate way to measure things because they were based on objects that varied in size. Recognizing the need for a global measurement system, the French Academy of Sciences developed a system in the late 1700s. Over the next 200 years, the metric system, now called the *International System of Units* (SI), was refined.

Using the International System of Units

Today, most scientists and other people in almost all countries use the International System of Units. One advantage of using SI measurements is that all scientists can share and compare their observations and results. Another advantage of the SI is that all units are based on the number 10, which is a number that is easy to use in calculations. **Table 1** contains the commonly used SI units.

✓ **Reading Check** Why was the International System of Units developed? (*See the Appendix for answers to Reading Checks.*)

READING WARM-UP

Objectives

● Explain the importance of the International System of Units.

● Determine appropriate units to use for particular measurements.

● Identify lab safety symbols, and determine what they mean.

Terms to Learn

meter	temperature
volume	area
mass	density

READING STRATEGY

Reading Organizer As you read this section, create an outline of the section. Use the headings from the section in your outline.

Table 1	Common SI Units and Conversions	
Length	**meter (m)**	
	kilometer (km)	1 km = 1,000 m
	decimeter (dm)	1 dm = 0.1 m
	centimeter (cm)	1 cm = 0.01 m
	millimeter (mm)	1 mm = 0.001 m
	micrometer (μm)	1 μm = 0.000001 m
	nanometer (nm)	1 nm = 0.000000001 m
Volume	**cubic meter (m³)**	
	cubic centimeter (cm³)	$1 \text{ cm}^3 = 0.000001 \text{ m}^3$
	liter (L)	$1 \text{ L} = 1 \text{ dm}^3 = 0.001 \text{ m}^3$
	milliliter (mL)	$1 \text{ mL} = 0.001 \text{ L} = 1 \text{ cm}^3$
Mass	**kilogram (kg)**	
	gram (g)	1 g = 0.001 kg
	milligram (mg)	1 mg = 0.000 001 kg
Temperature	**kelvin (K)**	
	Celsius (°C)	0°C = 273 K
		100°C = 373 K

Length

To measure length, a scientist uses meters (m). A **meter** is the basic SI unit of length. You may remember that SI units are based on the number 10. If you divide 1 m into 100 parts, for example, each part equals 1 cm. In other words, a centimeter equals one-hundredth of a meter. Some objects are so tiny that smaller units must be used. To describe the length of microscopic objects, scientists use micrometers (μm) or nanometers (nm). To describe the length of larger objects, scientists use kilometers (km).

Volume

Imagine that you are a scientist who needs to move some fossils to a museum. How many fossils will fit into a crate? The answer depends on the volume of the crate and the volume of each fossil. **Volume** is the measure of the size of a body or region in three-dimensional space.

The volume of a liquid is often given in liters (L). Liters are based on the meter. A cubic meter (1 m^3) is equal to 1,000 L. So, 1,000 L will fit into a box measuring 1 m on each side. The volume of a large, solid object is given in cubic meters. The volumes of smaller objects can be given in cubic centimeters (cm^3) or cubic millimeters (mm^3). To calculate the volume of a box-shaped object, multiply the object's length by its width and then multiply by its height.

The length, height, and width of irregularly shaped objects, such as rocks and fossils, are difficult to measure accurately. However, the volume of an irregularly shaped object can be determined by measuring the volume of liquid that the object displaces. The student in **Figure 1** is using a graduated cylinder to measure the volume of water a rock displaces.

meter the basic unit of length in the SI (symbol, m)

volume a measure of the size of a body or region in three-dimensional space

SCHOOL to HOME

Taking Measurements

With a parent, measure the width of your kitchen table, using your hands as a unit of measure. First, use your own hand to determine the width of the table. Then, have your parent use his or her hand to measure the width of the table. Compare your measurement with that of your parent. Was the number of units the same? Explain why it is important to use standard units of measurement.

ACTiViTY

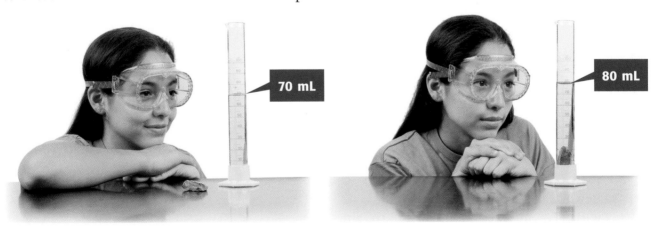

Figure 1 *This graduated cylinder contains 70 mL of water. After the rock was added, the water level moved to 80 mL. Because the rock displaced 10 mL of water and because 1 mL = 1 cm³, the volume of the rock is 10 cm³.*

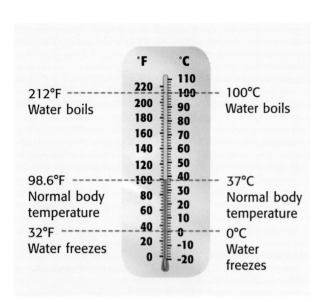

Figure 2 *This thermometer shows the relationship between degrees Fahrenheit and degrees Celsius.*

Mass

A measure of the amount of matter in an object is **mass.** The kilogram (kg) is the basic unit for mass. The kilogram is used to describe the mass of things such as boulders. But many common objects are not so large. Grams (g) are used to describe the mass of smaller objects, such as an apple. One thousand grams equals 1 kg. The mass of large objects, such as an elephant, is given in metric tons. A metric ton equals 1,000 kg.

Temperature

Temperature is the measure of how hot (or cold) something is. You are probably used to describing temperature with degrees Fahrenheit (°F). For example, if your body temperature is 101°F, you have a fever. Scientists, however, usually use degrees Celsius (°C). The thermometer in **Figure 2** shows the relationship between degrees Fahrenheit and degrees Celsius. The kelvin, the SI base unit for temperature, is also used in science.

mass a measure of the amount of matter in an object

temperature a measure of how hot (or cold) something is

area a measure of the size of a surface or a region

Area

How much paper would you need to cover the top of your desk? To answer this question, you must find the area of the desk. **Area** is a measure of how much surface an object has. The units for area are square units, such as square meters (m^2), square centimeters (cm^2), and square kilometers (km^2). To calculate the area of a square or rectangle, first measure the length and width. Then, use the following equation:

$$area = length \times width$$

Finding Area What is the area of a rectangle that has a length of 4 cm and a width of 5 cm?

Step 1: Write the equation for area.

area = length × width

Step 2: Replace the length and width with the measurements given in the problem, and solve.

area = 4 cm × 5 cm = 20 cm²

Now It's Your Turn

1. What is the area of a square whose sides measure 5 m?

2. What is the area of a book cover that is 22 cm wide and 28 cm long?

Density

The ratio of the mass of a substance to the volume of the substance is the substance's **density**. Because density is the ratio of mass to volume, units often used for density are grams per milliliter (g/mL) and grams per cubic centimeter (g/cm^3). You can calculate density by using the following equation:

$$density = \frac{mass}{volume}$$

Safety Rules!

Science is exciting and fun, but it can also be dangerous. Always follow your teacher's instructions. Before starting any scientific investigation, obtain your teacher's permission. Read the lab procedures completely and carefully before you start. Pay attention to safety information and caution statements. **Figure 3** shows the safety symbols that are used in this book.

Reading Check What should you do before you start a scientific investigation?

Figure 3 **Safety Symbols**

 Eye protection

 Clothing protection

 Hand safety

 Heating safety

 Electric safety

 Sharp object

 Chemical safety

Animal safety

Plant safety

density the ratio of the mass of a substance to the volume of the substance

SECTION Review

Summary

● The SI is the standard system of measurement used by scientists around the world.

● The basic SI units of measurement for length, volume, mass, and temperature are the meter, liter, kilogram, and kelvin, respectively.

● Safety rules must be followed at all times during a scientific investigation.

Using Key Terms

The statements below are false. For each statement, replace the underlined term to make a true statement.

1. The length multiplied by the width of an object is the <u>density</u> of the object.

2. The measure of the amount of matter in an object is the <u>area</u>.

Understanding Key Ideas

3. Which of the following SI units is most often used to measure length?
   a. meter
   b. liter
   c. gram
   d. degrees Celsius

4. What are two benefits of using the International System of Units?

5. At what temperature in degrees Celsius does water freeze?

6. Why is it important to follow safety rules?

Math Skills

7. Find the density of an object that has a mass of 34 g and a volume of 14 mL.

Critical Thinking

8. **Making Comparisons** Which weighs more: a pound of feathers or a pound of lead? Explain.

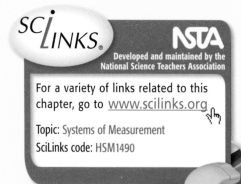

For a variety of links related to this chapter, go to www.scilinks.org

Topic: Systems of Measurement
SciLinks code: HSM1490

Model-Making Lab

Using Scientific Methods

Geologists often use a technique called core sampling to learn what underground rock layers look like. This technique involves drilling several holes in the ground in different places and taking samples of the underground rock or soil. Geologists then compare the samples from each hole at each depth to construct a diagram that shows the bigger picture.

In this activity, you will model the process geologists use to diagram underground rock layers. You will first use modeling clay to form a rock-layer model. You will then exchange models with a classmate, take core samples, and draw a diagram of your classmate's rock layers.

OBJECTIVES

Design a model to demonstrate core sampling.

Create a diagram of a classmate's model by using the core sample method.

MATERIALS

- knife, plastic
- modeling clay, three or four colors
- pan or box, opaque
- pencil, unsharpened
- pencils or markers, three or four colors
- PVC pipe, 1/2 in.

SAFETY

- Form a plan for your rock layers. Make a sketch of the layers. Your sketch should include the colors of clay in several layers of varying thicknesses. Note: Do not let the classmates who will be using your model see your plan.

- In the pan or box, mold the clay into the shape of the lowest layer in your sketch.

- Repeat the procedure described in the second bullet for each additional layer of clay. Exchange your rock-layer model with a classmate.

Ask a Question

1 Can unseen features be revealed by sampling parts of the whole?

Form a Hypothesis

2 Form a hypothesis about whether taking core samples from several locations will give a good indication of the entire hidden feature.

Test the Hypothesis

3 Choose three places on the surface of the clay to drill holes. The holes should be far apart and in a straight line. (Do not remove the clay from the pan or box.)

4 Slowly push the PVC pipe through all the layers of clay. Slowly remove the pipe.

5 Gently push the clay out of the pipe with an unsharpened pencil. This clay is a core sample.

6 Draw the core sample, and record your observations. Be sure to use a different color of pencil or marker for each layer.

7 Repeat steps 4–6 for the next two core samples. Make sure your drawings are side by side and in the same order as the samples in the model.

Analyze the Results

1 **Examining Data** Look at the pattern of rock layers in each of your core samples. Think about how the rock layers between the core samples might look. Then, make a diagram of the rock layers.

2 **Organizing Data** Complete your diagram by coloring the rest of each rock layer.

Draw Conclusions

3 **Evaluating Data** Use the plastic knife to cut the clay model along a line connecting the three holes. Remove one side of the model so that you can see the layers.

4 **Evaluating Models** How well does your rock-layer diagram match the model? Explain.

5 **Evaluating Methods** What are some limitations of your diagram as a model of the rock layers?

6 **Drawing Conclusions** Do your conclusions support your hypothesis? Explain your answer.

Applying Your Data

List two ways that the core-sampling method could be improved.

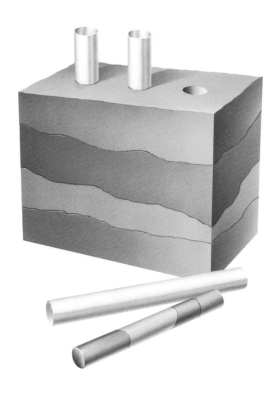

Chapter Review

USING KEY TERMS

Complete each of the following sentences by choosing the correct term from the word bank.

geology   astronomy

scientific methods   hypothesis

1 The study of the origin, history, and structure of the Earth and the processes that shape the Earth is called ___.

2 An explanation that is based on prior scientific research or observations and that can be tested is called a(n) ___.

3 ___ are a series of steps followed to solve problems.

UNDERSTANDING KEY IDEAS

Multiple Choice

4 The science that uses geology to study how humans affect the natural environment is

a. paleontology.

b. environmental science.

c. cartography.

d. volcanology.

5 A pencil measures 14 cm long. How many millimeters long is it?

a. 1.4 mm   **c.** 1,400 mm

b. 140 mm   **d.** 1,400,000 mm

6 Which of the following is NOT an SI unit?

a. meter   **c.** liter

b. foot   **d.** degrees Celsius

7 Which of the following is a limitation of models?

a. They are large enough to be seen.

b. They do not act exactly like the thing they model.

c. They are smaller than the thing they model.

d. They use familiar things to model unfamiliar things.

8 Gillette's hypothesis was

a. supported by his results.

b. not supported by his results.

c. based only on observations.

d. based only on what he already knew.

Short Answer

9 Why would scientific investigations lead to new scientific investigations?

10 How and why do scientists use models?

11 What are three types of models? Give an example of each.

12 What problems could occur if scientists didn't communicate the results of their investigations?

13 What problems could occur if there were not an International System of Units?

14 Which safety symbols would you expect to see for an experiment that requires the use of acid?

CRITICAL THINKING

15 **Concept Mapping** Use the following terms to create a concept map: *Earth science, scientific methods, hypothesis, problem, question, experiment,* and *observations.*

16 **Analyzing Processes** Why do you not need to complete the steps of scientific methods in a specific order?

17 **Evaluating Conclusions** Why might two scientists working on the same problem draw different conclusions?

18 **Analyzing Methods** Scientific methods often begin with observation. How does observation limit what scientists can study?

19 **Making Comparisons** A rock that contains fossil seashells might be studied by scientists in at least two branches of Earth science. Name those branches. Why did you choose those two branches?

INTERPRETING GRAPHICS

Use the graph below to answer the questions that follow.

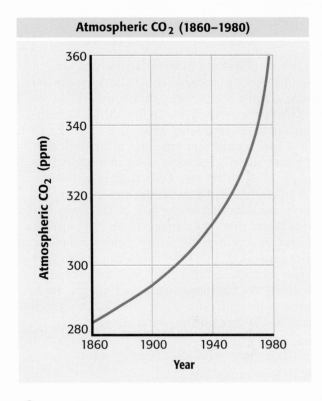

Atmospheric CO$_2$ (1860–1980)

20 Has the amount of CO$_2$ in the atmosphere increased or decreased since 1860?

21 The line on the graph is curved. What does this curve indicate?

22 Was the rate of change in the level of CO$_2$ between 1940 and 1960 higher or lower than it was between 1880 and 1900? How can you tell?

23 What conclusions can you draw from reading this graph?

READING

Read each of the passages below. Then, answer the questions that follow each passage.

Passage 1 Scientists look for answers by asking questions. For instance, scientists have wondered if there is some relationship between Earth's core and Earth's magnetic field. To form their hypothesis, scientists started with what they knew: Earth has a dense, solid inner core and a molten outer core. They then created a computer model to <u>simulate</u> how Earth's magnetic field is generated. The model predicted that Earth's inner core spins in the same direction as the rest of the Earth but slightly faster than the surface. If that hypothesis is correct, it might explain how Earth's magnetic field is generated. But how could the researchers test the hypothesis? Because scientists couldn't drill down to the core, they had to get their information indirectly. They decided to track seismic waves created by earthquakes.

1. In the passage, what does *simulate* mean?
  A to look or act like
  B to process
  C to calculate
  D to predict

2. According to the passage, what do scientists wonder?
  F if the Earth's inner core was molten
  G if there was a relationship between Earth's core and Earth's magnetic field
  H if the Earth had a solid outer core
  I if computers could model the Earth's core

3. What did the model predict?
  A The Earth's outer core is molten.
  B The Earth's inner core is molten.
  C The Earth's inner core spins in the same direction as the rest of the Earth.
  D The Earth's outer core spins in the same direction as the rest of the Earth.

Passage 2 Scientists analyzed seismic data for a 30-year period. They knew that seismic waves traveling through the inner core along a north-south path travel faster than waves passing through it along an east-west line. Scientists searched seismic data records to see if the <u>orientation</u> of the "fast path" for seismic waves changed over time. They found that in the last 30 years, the direction of the "fast path" for seismic waves had indeed shifted. This is strong evidence that Earth's core does travel faster than the surface, and it strengthens the hypothesis that the spinning core creates Earth's magnetic field.

1. In the passage, what does *orientation* mean?
  A speed
  B direction
  C magnetic field
  D intensity

2. What evidence did scientists find?
  F The Earth's core does travel faster than the surface.
  G The "fast path" does not change.
  H Seismic waves travel faster along an east-west line.
  I The spinning core does not create the Earth's magnetic field.

3. What do scientists hypothesize about the Earth's magnetic field?
  A It was found in the last 30 years.
  B It travels faster along a north-south path.
  C It is losing its strength.
  D It is created by the spinning core.

The table below contains data that shows the relationship between volume and pressure. Use the table to answer the questions that follow.

Volume (L)	Pressure (kPa)
0.5	4,960
1.0	2,480
2.0	1,240
3.0	827

1. What is the pressure when the volume is 2.0 L?

  A 4,960 kPa

  B 2,480 kPa

  C 1,240 kPa

  D 827 kPa

2. What is the volume when the pressure is 827 kPa?

  F 0.5 L

  G 1.0 L

  H 2.0 L

  I 3.0 L

3. What is the change in pressure when the volume is increased from 0.5 L to 1.0 L?

  A 4,960 kPa

  B 2,480 kPa

  C 1,240 kPa

  D 0.50 kPa

4. Which of the following patterns best describes the data?

  F When the volume is doubled, the pressure is tripled.

  G When the volume is tripled, the pressure is cut in half.

  H As the volume increases, the pressure remains the same.

  I As the volume increases, the pressure decreases.

Read each question below, and choose the best answer.

1. The original design for a boat shows a rectangular shape that is 5 m long and 1.5 m wide. If the design is reduced to 3.4 m long and 1 m wide, by how much does the area of the boat decrease?

  A 1.7 m²

  B 4.1 m²

  C 7.5 m²

  D 9.2 m²

2. If *density = mass/volume,* what is the density of an object that has a mass of 50 g and a volume of 2.6 cm³?

  F 0.052 cm³/g

  G 19.2 g/cm³

  H 47.4 g/cm³

  I 130 g/cm³

3. During a chemical change, two separate pieces of matter combined into one. The mass of the final product is 82 g. The masses of the original pieces must equal the final product's mass. What are the possible masses of the original pieces of matter?

  A 2 g and 18 g

  B 2 g and 41 g

  C 12 g and 8 g

  D 42 g and 40 g

4. An adult *Seismosaurus hallorum* weighs 82 tons. A baby *Seismosaurus hallorum* weighs 46 tons. The weight of the baby *Seismosaurus hallorum* is what percentage of the weight of the adult *Seismosaurus hallorum*?

  F 24%

  G 44%

  H 56%

  I 98%

Standardized Test Preparation   **61**

Science in Action

Science, Technology, and Society

A "Ship" That Flips?

Does your school's laboratory have doors on the floor or tables bolted sideways to the walls? A lab like this exists, and you can find it floating in the ocean. *FLIP*, or *Floating Instrument Platform*, is a 108 m long ocean research vessel that can tilt 90°. *FLIP* is towed to an area that scientists want to study. To flip the vessel, empty chambers within the vessel are filled with water. The *FLIP* begins tilting until almost all of the vessel is underwater. Having most of the vessel below the ocean's surface stabilizes the vessel against wind and waves. Scientists can collect accurate data from the ocean, even during a hurricane!

Social Studies ACTiViTY

Design your own *FLIP*. Make a map on poster board. Draw the layout of a living room, bathroom, and bedroom before your *FLIP* is tilted 90°. Include entrances and walkways to use when *FLIP* is not flipped.

Weird Science

It's Raining Fish and Frogs

What forms of precipitation have you seen fall from the sky? Rain, snow, hail, sleet, or fish? Wait a minute! Fish? Fish and frogs might not be a form of precipitation, but as early as the second century, they have been reported to fall from the sky during rainstorms. Scientists theorize that tornadoes or waterspouts that suck water into clouds can also suck up unsuspecting fish, frogs, or tadpoles that are near the surface of the water. After being sucked up into the clouds and carried a few miles, these reluctant travelers then rain down from the sky.

Language Arts ACTiViTY

WRITING SKILL You are a reporter for your local newspaper. On a rainy day in spring, while driving to work, you witness a downpour of frogs and fish. You pull off to the side of the road and interview other witnesses. Write an article describing this event for your local newspaper.

Careers

Sue Hendrickson

Paleontologist Could you imagine having a job in which you spent all day digging in the dirt? This is just one of Sue Hendrickson's job descriptions. But Hendrickson does not dig up flowers. Hendrickson is a paleontologist, and she digs up dinosaurs! Her most famous discovery is the bones of a *Tyrannosaurus rex*. *T. rex* is one of the largest meat-eating dinosaurs. It lived between 65 million and 85 million years ago. Walking tall at 6 m, *T. rex* was approximately 12.4 m long and weighed between 5 and 7 tons. Hendrickson's discovery is the most complete set of bones ever found of the *T. rex*. The dinosaur was named Sue to honor Hendrickson for her important find. From these bones, Hendrickson and other scientists have been able to learn more about the dinosaur, including how it lived millions of years ago. For example, Hendrickson and her team of scientists found the remains of Sue's last meal, part of a a duck-billed, plant-eating dinosaur called *Edmontosaurus* that weighed approximately 3.5 tons!

Math Activity

ACTiViTY

The *T. rex* named Sue weighed 7 tons and the *Edmontosaurus* weighed 3.5 tons. How much smaller is *Edmontosaurus* than Sue? Express your answer as a percentage.

To learn more about these Science in Action topics, visit **go.hrw.com** and type in the keyword **HZ5WESF.**

Current Science

Check out Current Science® articles related to this chapter by visiting go.hrw.com. Just type in the keyword HZ5CS01.

The World of Physical Science

About the PHOTO

Flippers work great to help penguins move through the water. But could flippers help ships, too? Two scientists have been trying to find out. By using scientific methods, they are asking questions such as, "Would flippers use less energy than propellers do?" As a result of these investigations, ships may have flippers like those of penguins someday!

PRE-READING ACTiViTY

**Graphic
Organizer**

Spider Map Before you read the chapter, create the graphic organizer entitled "Spider Map" described in the **Study Skills** section of the Appendix. Label the circle "Scientific Models." Create a leg for each type of scientific model. As you read the chapter, fill in the map with details about each type of scientific model.

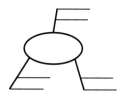

Figure It Out

In this activity, you will make observations and use them to solve a puzzle, just as scientists do.

Procedure

1. Get the **five shapes** shown here from your teacher.

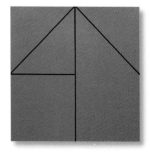

2. Observe the drawing at right. Predict how the five shapes could be arranged to make the fish.

3. Test your idea. You may have to try several times. (Hint: Shapes can be turned over.)

Analysis

1. Did you solve the puzzle just by making observations? What observations helped the most?

2. How did testing your ideas help?

Exploring Physical Science

You're eating breakfast. You look down and notice that your reflection in your spoon is upside down! You wonder, Why is my reflection upside down even though I'm holding the spoon right side up?

Congratulations! You just completed the first steps toward becoming a scientist. How did you do it? You observed the world around you. Then, you asked questions about your observations. That's what most of science is all about.

READING WARM-UP

Objectives

● Explain that science involves asking questions.

● Describe the relationship of matter and energy to physical science.

● Describe the two branches of physical science.

● Identify three areas of science that use physical science.

Terms to Learn

science
physical science

READING STRATEGY

Prediction Guide Before reading this section, write the title of each heading in this section. Next, under each heading, write what you think you will learn.

That's Science!

Science is a process of gathering knowledge about the natural world. Asking a question is often the first step in the process of gathering knowledge. The world around you is full of amazing things that can lead you to ask questions. The student in **Figure 1** didn't have to look very far to realize that she had some questions to ask.

☑ **Reading Check** What is often the first step in gathering knowledge? (*See the Appendix for answers to Reading Checks.*)

Everyday Science

Everyday actions such as timing the microwave popcorn and using the brakes on your bicycle use your knowledge of science. You learned how to do these things by making observations and asking questions. Making observations and asking questions is what science is all about. And because science is all around, you might not be surprised to learn that there are many branches of science. Physical science is the science you will learn about in this book. But physical science is just one of the many different branches of science.

Why can I see a reflection in a spoon?

What causes high and low tides?

Why do I feel pain when I stub my toe?

Figure 1 *Part of science is asking questions about the world around you.*

Figure 2 *The cheetah, the fastest land mammal, uses a lot of energy when running full speed. But a successful hunt will supply the energy the cheetah needs to live.*

What Is Physical Science?

Physical science is the study of matter and energy. *Matter* is the "stuff" that everything is made of. Even stuff so small that you can't see it is matter. Your shoes, your pencil, and even the air you breathe are made of matter. And all of this matter has energy. *Energy* is the ability to do work. But energy is easier to describe than to explain. For example, energy is partly responsible for rainbows in the sky. But energy isn't the rainbow itself. All moving objects have energy of motion, such as the cheetah shown in **Figure 2.**

Food also has energy. When you eat food, the energy in the food is transferred to you. You can use that energy to carry out your daily activities. But energy isn't always associated with motion or food. All matter, including matter that isn't moving, has energy. So, because the baseball in **Figure 3** is matter, it has energy even though it is not yet moving.

science the knowledge obtained by observing natural events and conditions in order to discover facts and formulate laws or principles that can be verified or tested

physical science the scientific study of nonliving matter

A Study of Matter and Energy

As you explore physical science, you'll learn more about matter and energy. And you will see how matter and energy relate to each other. For example, both paper and gold are matter. But why will paper burn, and gold will not? And why is throwing a bowling ball harder than throwing a baseball? How can water turn into steam and back into water? All of the answers to these questions have to do with matter and energy. It is hard to talk about matter without talking about energy. However, sometimes it is useful to focus on one or the other. Physical science is also often divided into two categories: chemistry and physics.

Figure 3 *The baseball has energy even before the boy throws it, because it is matter, and all matter has energy.*

Branches of Physical Science

Physical science is usually divided into chemistry and physics. But both chemistry and physics can be further broken down into many more specialized areas of study. For example, chemistry includes organic chemistry, which is the study of substances made of carbon. And geophysics, one of the branches of physics, includes the study of the vibrations deep inside the Earth that are caused by earthquakes.

Chemistry—A Matter of Reactions!

Chemistry is the study of all forms of matter, including how matter interacts with other matter. Chemistry looks at the structure and properties of matter. For example, some substances behave one way under high temperature and high pressure. But other substances will behave very differently under the same conditions. The scientist in **Figure 4** is studying the properties of different kinds of materials. He is trying to find materials that have unusual properties, such as the ability to withstand very high heat.

Chemistry is also the study of how substances change. A chemical reaction takes place when one substance reacts with another substance to make a new substance. Chemical reactions are taking place around you all of the time. When your body digests food, a chemical reaction is taking place. Chemical reactions are needed when you take a photo of your best friend, when your parent starts the car engine, and when you turn on a flashlight.

✓ **Reading Check** What are three things that chemistry studies?

Figure 4 *A materials scientist uses his knowledge of chemistry to study the properties of different kinds of substances.*

Figure 5 *When you study physics, you'll learn how energy causes the motion that makes a roller-coaster ride so exciting.*

Physics—A Matter of Energy

Like chemistry, physics deals with matter. But physics looks mostly at energy and the way that energy affects matter. Studying different forms of energy is what studying physics is all about. Energy can make matter do some interesting things. For example, have you ever wondered what keeps a roller coaster, such as the one shown in **Figure 5,** on its tracks? The study of physics will help answer this question.

Motion, force, gravity, electricity, light, and heat are parts of physics. They are also things that you experience in your daily life. For example, if you have ever ridden a bike, you are aware that force causes motion. If you have ever used a compass, you have dealt with the concept of magnetism. Do you know why you see a rainbow after a rainstorm? Or, do you know why shifting gears on your bicycle makes it easier to pedal? You will learn the answers to these questions, as well as many others, as you study physical science.

INTERNET ACTIVITY

For another activity related to this chapter, go to **go.hrw.com** and type in the keyword **HP5WPSW.**

CONNECTION TO Environmental Science

Thermal Pollution Factories are often built along the banks of rivers. The factories use the river water to cool the engines of their machinery. Then, the hot water is poured back into the river. Energy, in the form of heat, is transferred from this water to the river water. The increase in temperature results in the death of many living things. Research how thermal pollution causes fish to die. Also, find out what many factories are doing to prevent thermal pollution. Make a brochure that explains what thermal pollution is and what is being done to prevent it.

ACTIVITY

Physical Science: All Around You

Believe it or not, matter and energy are not just concepts in physical science. What you learn about matter and energy is important for other science classes, too.

Meteorology

The study of Earth's atmosphere, especially in relation to weather and climate, is called *meteorology* (MEET ee uhr AHL uh jee). A *meteorologist* (MEET ee uhr AHL uh jist) is a person who studies the atmosphere. One of the most common careers meteorologists have is weather forecasting. But other meteorologists specialize in and even chase tornadoes! These meteorologists predict where a tornado is likely to form. Then, they drive very close to the site to gather data, as shown in **Figure 6.** By gathering data this way, scientists hope to understand tornadoes better. Meteorologists must have knowledge of physical science. They must understand high and low pressure, motion, and force before they can predict how tornadoes will behave.

Geology

The study of the origin, history, and structure of Earth is called *geology*. Some geologists are geochemists (JEE oh KEM ists). A *geochemist* is a person who specializes in the chemistry of rocks, minerals, and soil. Geochemists, such as the one in **Figure 7,** try to find out what the environment was like when these materials formed and what has happened to the materials since they formed. To understand how rocks and soil have changed over millions of years, a geochemist must have a knowledge of heat, force, and chemistry.

Reading Check What does a geochemist study?

Figure 6 *These meteorologists are risking their lives to gather data about tornadoes.*

Figure 7 *This geochemist takes rock samples from the field. Then, she studies them in her laboratory.*

Biology

Students are often surprised that life science and physical science are related. But chemistry and physics explain many things that happen in biology. For example, a chemical reaction explains how animals, such as the cow in **Figure 8,** get energy from food. Sugar, $C_6H_{12}O_6$, which is produced by the plant, reacts with oxygen. As a result, carbon dioxide, water, and energy are produced. This reaction can be shown by the following chemical equation:

$$C_6H_{12}O_6 + 6O_2 \rightarrow 6CO_2 + 6H_2O + energy$$

Figure 8 *The cow gets energy by eating grass and other foods that contain sugars.*

SECTION Review

Summary

- Science is a process of gathering knowledge about the natural world.
- Physical science is the study of matter and energy.
- Physical science is divided into the study of physics and chemistry.
- Chemistry studies the structure and properties of matter and how matter changes.
- Physics looks at energy and the way that energy affects matter.
- A knowledge of physical science is important for many areas of science, such as geology and biology.

Using Key Terms

1. In your own words, write a definition for each of the following terms: *science* and *physical science*.

Understanding Key Ideas

2. Which of the following statements is true?
   a. Energy is the ability to do work.
   b. Air is made of matter.
   c. All matter has energy.
   d. All of the above

3. What are three areas of science that rely on physical science?

4. What is the difference between chemistry and physics?

Math Skills

5. You want to know which month had the highest percentage of rainy days for your city last year. Your investigation gave the following results: March had 5 days of rain, April had 8 days of rain, and May had 3 days of rain. For each month, what percentage of the month had rainy days? Which month had the highest percentage of rainy days?

Critical Thinking

6. **Applying Concepts** How do you think science is used by a pharmacist? by a firefighter?

7. **Analyzing Ideas** You are building a go-cart and want to know how to make it go as fast as possible. Which branch, or branches, of science would you study? Explain your answer.

8. **Identifying Relationships** Describe three things that you do every day that use your experience with physical science.

9. **Making Inferences** Botany is the study of plants. What role do you think physical science plays in botany?

Developed and maintained by the National Science Teachers Association

For a variety of links related to this chapter, go to www.scilinks.org

Topic: Scientific Inquiry; Careers in Science
SciLinks code: HSM1357; HSM0225

Scientific Methods

Imagine that you are trying to improve ships. Would you study the history of shipbuilding? Would you investigate different types of fuel? Would you observe animals that move easily through water, such as dolphins and penguins?

Two scientists from the Massachusetts Institute of Technology (MIT) thought that studying penguins was a great way to improve ships! James Czarnowski (zahr NOW SKEE) and Michael Triantafyllou (tree AHN ti FEE loo) used scientific methods to develop *Proteus* (PROH tee uhs), the penguin boat. In the next few pages, you will learn how these scientists used scientific methods to answer their questions.

What Are Scientific Methods?

Scientific methods are the ways in which scientists answer questions and solve problems. As scientists look for answers, they often use the same steps. But there is more than one way to use the steps. Look at **Figure 1.** This figure is an outline of the six steps on which scientific methods are based. Scientists may use all of the steps or just some of the steps during an investigation. They may even repeat some of the steps or do the steps in a different order. How they choose to use the steps depends on what works best to answer their question.

READING WARM-UP

Objectives

● Explain what scientific methods are.

● Explain how scientific methods are used to answer questions.

● Describe how a hypothesis is formed and tested.

● Identify methods that are used to analyze data.

● Explain how a conclusion can support or disprove a hypothesis.

● List methods of communicating data.

Terms to Learn

scientific methods
observation
hypothesis
data

READING STRATEGY

Reading Organizer As you read this section, make a flowchart of the steps used in scientific methods.

scientific methods a series of steps followed to solve problems

Figure 1 Steps of Scientific Methods

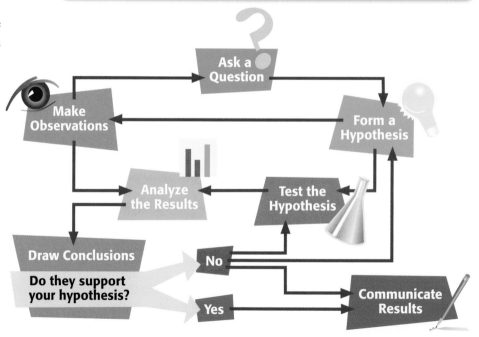

Ask a Question

Make Observations

Form a Hypothesis

Analyze the Results

Test the Hypothesis

Draw Conclusions

Do they support your hypothesis?

No

Yes

Communicate Results

Asking a Question

Asking a question helps focus the purpose of an investigation. Scientists often ask a question after making many observations. **Observation** is any use of the senses to gather information. Noting that the sky is blue or that a cotton ball feels soft is an observation. Measurements are observations that are made with tools, such as the ones shown in **Figure 2.** Keep in mind that observations can be made (and should be accurately recorded) at any point during an investigation.

✓ **Reading Check** What is the purpose of asking questions? (*See the Appendix for answers to Reading Checks.*)

A Real-World Question

Czarnowski and Triantafyllou, shown in **Figure 3,** are engineers, scientists who put scientific knowledge to practical use. Czarnowski was a graduate student at the Massachusetts Institute of Technology. He and Triantafyllou, his professor, worked together to observe boat propulsion (proh PUHL shuhn) systems. Then, they investigated how to make these systems work better. A propulsion system is what makes a boat move. Most boats have propellers to move them through the water.

Czarnowski and Triantafyllou studied the efficiency (e FISH uhn see) of boat propulsion systems. *Efficiency* compares energy output (the energy used to move the boat forward) with energy input (the energy supplied by the boat's engine). From their observations, Czarnowski and Triantafyllou learned that boat propellers are not very efficient.

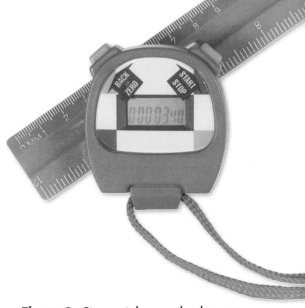

Figure 2 *Stopwatches and rulers are among the many tools used to make observations.*

observation the process of obtaining information by using the senses

Figure 3 *James Czarnowski* (left) *and Michael Triantafyllou* (right) *made observations about how boats work in order to develop* Proteus.

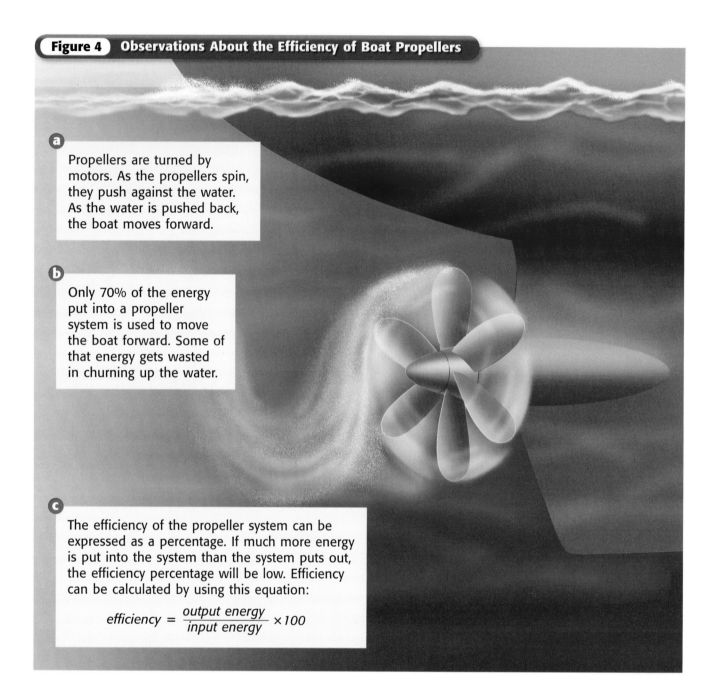

Figure 4 Observations About the Efficiency of Boat Propellers

a Propellers are turned by motors. As the propellers spin, they push against the water. As the water is pushed back, the boat moves forward.

b Only 70% of the energy put into a propeller system is used to move the boat forward. Some of that energy gets wasted in churning up the water.

c The efficiency of the propeller system can be expressed as a percentage. If much more energy is put into the system than the system puts out, the efficiency percentage will be low. Efficiency can be calculated by using this equation:

$$efficiency = \frac{output\ energy}{input\ energy} \times 100$$

The Importance of Boat Efficiency

Most boats that have propellers, shown in **Figure 4,** are only about 70% efficient. But is boat efficiency important, and if so, why? Yes, boat efficiency is important because it saves many resources. Making only a small fraction of U.S. boats and ships just 10% more efficient would save millions of liters of fuel per year. Saving fuel means saving money. It also means using less of Earth's supply of fossil fuels. Based on their observations and all of this information, Czarnowski and Triantafyllou were ready to ask a question: How can boat propulsion systems be made more efficient?

✓ **Reading Check** Why is boat efficiency important?

Figure 5 *Penguins use their flippers to "fly" underwater. As they pull their flippers toward their body, they push against the water, which propels them forward.*

Forming a Hypothesis

Once you've asked your question and made observations, you are ready to form a hypothesis (hie PAHTH uh sis). A **hypothesis** is a possible explanation or answer to a question. You can use what you already know and what you have observed to form a hypothesis.

A good hypothesis is testable. In other words, information can be gathered or an experiment can be designed to test the hypothesis. A hypothesis that is not testable isn't necessarily wrong. But there is no way to show whether the hypothesis is right or wrong.

hypothesis an explanation that is based on prior scientific research or observations and that can be tested

A Possible Answer from Nature

Czarnowski and Triantafyllou wanted to base their hypothesis on an example from nature. Czarnowski had made observations of penguins swimming at the New England Aquarium. He observed how quickly and easily the penguins moved through the water. **Figure 5** shows how penguins propel themselves. Czarnowski also observed that penguins, like boats, have a rigid body. These observations led to a hypothesis: A propulsion system that imitates the way that a penguin swims will be more efficient than a propulsion system that uses propellers.

Making Predictions

Before scientists test a hypothesis, they often predict what they think will happen when they test the hypothesis. Scientists usually state predictions in an if-then statement. The engineers at MIT might have made the following prediction: *If* two flippers are attached to a boat, *then* the boat will be more efficient than a boat powered by propellers.

CONNECTION TO Biology

Adaptations Penguins, though flightless, are better adapted to water and extreme cold than any other birds are. Research these amazing birds to learn how they are adapted to their environment. Also, investigate the speed at which penguins can swim. Present this information in a poster.

That's Swingin'!

1. Make a pendulum. Tie a **piece of string** to a **ring stand.** Hang a **small weight** from the string.

2. Form a testable hypothesis about one factor (such as the mass of the small weight) that may affect the rate at which the pendulum swings.

3. Predict the results as you change this factor (the variable).

4. Test your hypothesis. Record the number of swings made in 10 s for each trial.

5. Was your hypothesis supported? Analyze your results.

Testing the Hypothesis

After you form a hypothesis, you must test it. You must find out if it is a reasonable answer to your question. Testing helps you find out if your hypothesis is pointing you in the right direction or is way off the mark. If your hypothesis is way off the mark, you may have to change it.

Controlled Experiments

One way to test a hypothesis is to do a controlled experiment. A *controlled experiment* compares the results from a control group with the results from experimental groups. The groups are the same except for one factor. This factor is called a *variable.* The results of the experiment will show the effect of the variable.

Sometimes, a controlled experiment is not possible. Stars, for example, are too far away to be used in an experiment. In such cases, you can make more observations or do research. Another investigation may require you to make or build a device to test. You can then test your device to see if it does what you expected it to do and if the results support your hypothesis. Czarnowski and Triantafyllou did such a controlled experiment. They built *Proteus,* the penguin boat, shown in **Figure 6.**

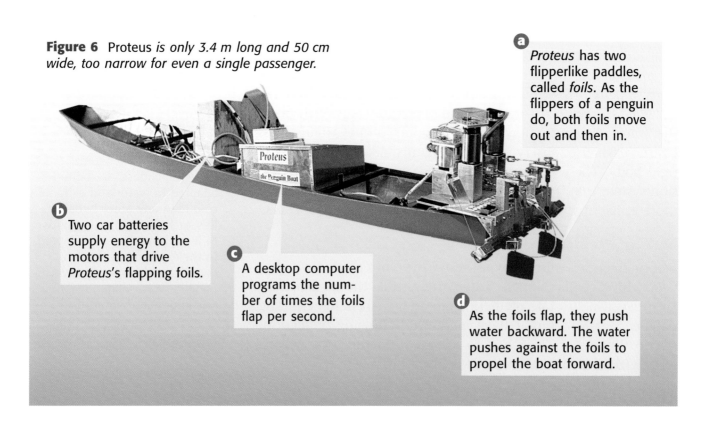

Figure 6 Proteus *is only 3.4 m long and 50 cm wide, too narrow for even a single passenger.*

a *Proteus* has two flipperlike paddles, called *foils.* As the flippers of a penguin do, both foils move out and then in.

b Two car batteries supply energy to the motors that drive *Proteus*'s flapping foils.

c A desktop computer programs the number of times the foils flap per second.

d As the foils flap, they push water backward. The water pushes against the foils to propel the boat forward.

Figure 7 Graphs of the Test Results

This line graph shows that *Proteus* was most efficient when its foils were flapping about 1.7 times per second.

This bar graph shows that *Proteus* is 17% more efficient than a propeller-driven boat.

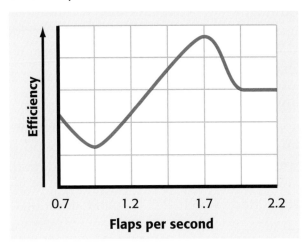

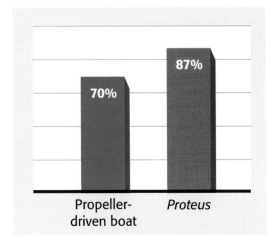

Testing *Proteus*

Czarnowski and Triantafyllou took *Proteus* out into the open water of the Charles River in Boston when they were ready to collect data. **Data** are any pieces of information acquired through experimentation. For each test, data such as the flapping rate, the energy used by the motors, and the speed achieved by the boat were carefully recorded. But the only factor the scientists changed was the flapping rate. The input energy was determined by how much energy was used. The output energy was determined from the speed *Proteus* reached.

data any pieces of information acquired through observation or experimentation

Analyzing the Results

After you collect and record your data, you must analyze them. You must find out if the results of your test support the hypothesis. Sometimes, doing calculations can help you learn more about your results. Organizing data into tables and graphs makes relationships between information easier to see.

Reading Check Why are graphs and charts useful for analyzing results?

Analyzing *Proteus*

Czarnowski and Triantafyllou used the data for input energy and output energy to calculate *Proteus*'s efficiency for different flapping rates. These data are graphed in **Figure 7.** The scientists compared *Proteus*'s highest level of efficiency with the average efficiency of a propeller-driven boat. As you can see, the data support the scientists' hypothesis that penguin propulsion is more efficient than propeller propulsion.

Drawing Conclusions

At the end of an investigation, you must draw a conclusion. You could conclude that your results support your hypothesis. Or you could conclude that your results do *not* support your hypothesis. Or you might even conclude that you need more information. Your conclusion can help guide you in deciding what to do next. You could ask new questions, gather more information, or change the procedure.

The *Proteus* Conclusion

After Czarnowski and Triantafyllou analyzed the results of their test, they ran many more trials. Again, they found that the penguin propulsion system was more efficient than a propeller propulsion system. So, they concluded that their hypothesis was supported, which led to more questions, as **Figure 8** shows.

Communicating Results

One of the most important steps in any investigation is to communicate your results. You can write a scientific paper, make a presentation, or create a Web site. Telling others what you learned keeps science going. Other scientists can then conduct their own tests.

✓ Reading Check What are some ways to communicate the results of an investigation?

Communicating About *Proteus*

Czarnowski and Triantafyllou published their results in academic papers, science magazines, and newspapers. They also displayed the results of their project on the Internet. These reports allow other scientists to conduct additional research about *Proteus*.

CONNECTION TO Language Arts

WRITING SKILL Communicating Without Words

Research methods of non-verbal communication. In the past, how did people communicate with each other without talking? Write a one-page essay describing what you have learned about nonverbal communication.

Figure 8 *Can a penguin propulsion system be used on large ships, such as an oil tanker? The research continues!*

Summary

- Scientific methods are the ways in which scientists answer questions and solve problems.
- Asking a question usually results from making an observation. Questioning is often the first step of using scientific methods.
- A hypothesis is a possible explanation or answer to a question. A good hypothesis is testable.

- After testing a hypothesis, you should analyze your results. Analyzing is usually done by using calculations, tables, and graphs.
- After analyzing your results, you should draw conclusions about whether your hypothesis is supported.
- Communicating your results allows others to check or continue your work. You can communicate through reports, posters, and the Internet.

Using Key Terms

The statements below are false. For each statement, replace the underlined term to make a true statement.

1. <u>Observations</u> are the ways in which scientists answer questions and solve problems.

2. <u>Hypotheses</u> are pieces of information that are gathered through experimentation.

3. <u>Data</u> are possible explanations or answers to a question.

Understanding Key Ideas

4. A controlled experiment
   a. is not always possible.
   b. contains a test group.
   c. has only one variable.
   d. All of the above

5. Name the steps that can be used in scientific methods.

Critical Thinking

6. **Analyzing Methods** Explain how the accuracy of your observations might affect how you develop a hypothesis.

7. **Applying Concepts** You want to test different shapes of kites to see which shape results in the strongest lift, or upward force, in the air. List some factors that need to be the same for each trial so that the only variable is the shape of the kite.

Interpreting Graphics

Use the graph below to answer the questions that follow.

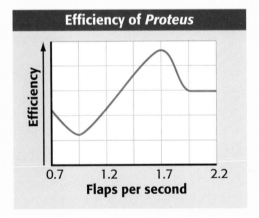

Efficiency of *Proteus*

Efficiency (vertical axis)

Flaps per second (horizontal axis): 0.7, 1.2, 1.7, 2.2

8. At what number of flaps per second is *Proteus* most efficient? least efficient?

9. At approximately what point does the efficiency appear neither to increase nor to decrease?

SCiLINKS®

NSTA
Developed and maintained by the National Science Teachers Association

For a variety of links related to this chapter, go to www.scilinks.org

Topic: Scientific Methods
SciLinks code: HSM1359

Scientific Models

How much like a penguin is Proteus? Proteus *doesn't have feathers and isn't a living thing. But its "flippers" create the same kind of motion that a penguin's flippers do.*

The MIT engineers built *Proteus* to mimic the way a penguin swims so that they could gain a greater understanding about boat propulsion. In other words, they made a model.

Models in Science

A **model** is a representation of an object or system. A model uses something familiar to help you understand something that is not familiar. For example, models of human body systems can help you understand how the body works. Models can also be used to explain the past and the present. They can even be used to predict future events. There are three common kinds of scientific models. They are physical, mathematical, and conceptual (kuhn SEP choo uhl) models. However, models have limitations because they are never exactly like the real thing.

READING WARM-UP

Objectives

● Explain how models represent the natural world.

● Identify three types of models used in science.

● Describe theories and laws.

Terms to Learn

model
theory
law

READING STRATEGY

Discussion Read this section silently. Write down questions that you have about this section. Discuss your questions in a small group.

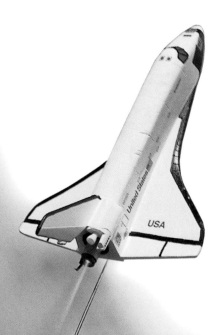

Figure 1 *Using a model of a space shuttle can help you understand how a real space shuttle works.*

Physical Models

Model airplanes, dolls, and drawings are examples of physical models. A model of a molecule can show you the shape of the molecule, which you cannot see. But this model wouldn't let you see how the molecule interacts with other molecules. Other kinds of physical models can help you understand certain concepts. For example, look at the model space shuttle and the real space shuttle in **Figure 1.** Launching a model like the one on the left can help you understand how a real space shuttle blasts off into space.

Mathematical Models

Every day, people try to predict the weather. One way to predict the weather is to use mathematical models. A mathematical model is made up of mathematical equations and data. Some mathematical models are simple. These models allow you to calculate things such as forces and acceleration. But other mathematical models are so complex that only computers can handle them. Some of these very complex models have many variables. Sometimes, certain variables that no one thought of exist in a model. A change in any variable could cause the model to fail.

Reading Check Name a possible limitation of a mathematical model. (*See the Appendix for answers to Reading Checks.*)

Conceptual Models

The third kind of model is a conceptual model. Some conceptual models are systems of ideas. Others are based on making comparisons with familiar things to help illustrate or explain an idea. The big bang theory is a conceptual model that describes how the planets and galaxies formed. This model is described in **Figure 2.** Although the big bang theory is widely accepted by astronomers, some data do not fit the model. For example, scientists have calculated the ages of some old, nearby stars. If the calculations are right, some of these stars are older than the universe itself. So, conceptual models may not take certain data into account. Or the models may rely on certain ideas but not on others.

model a pattern, plan, representation, or description designed to show the structure or workings of an object, system, or concept

Weather Forecasting
Watch the weather forecast on TV. You will see several models that a weather reporter uses to inform you about the weather in your area. In your **science journal,** describe two of these models and explain how each model is used to represent the weather. Describe some of the advantages and disadvantages of each model.

Figure 2 *The big bang theory says that 12 billion to 15 billion years ago, an event called the* big bang *sent matter in all directions. This matter eventually formed the galaxies and planets.*

Models: The Right Size

Models are often used to represent things that are very small or very large. Some particles of matter are too small to see. The Earth and the solar system are too large to see completely. So, a model can help you picture the thing in your mind. Sometimes, models are used to learn about things you cannot see, such as sound waves. Look at **Figure 3.** A coiled spring toy is often used as a model of sound waves because the spring toy behaves similar to the way sound waves do.

Using Models to Build Scientific Knowledge

Models not only can represent scientific ideas and objects but also can be tools that are useful to help you learn new information.

Scientific Theories

Models are often used to help illustrate and explain scientific theories. In science, a **theory** is an explanation for many hypotheses and observations. Usually, these hypotheses have been supported by repeated tests. A theory not only explains an observation you've made but also can predict what might happen in the future.

Scientists use models to help guide their search for new information. This information can help support a theory or can show that the theory is wrong. Keep in mind that models can be changed or replaced. These changes happen when scientists make new observations. Because of these new observations, scientists may have to change their theories. **Figure 4** compares an old model with a current model.

✓ **Reading Check** What two things can a theory explain?

Figure 3 *The compressed coils on the spring toy can be used to model the way air particles are crowded together in a sound wave.*

theory an explanation that ties together many hypotheses and observations

law a summary of many experimental results and observations; a law tells how things work

Figure 4 *These models show how scientists' idea of the atom has changed over time as new information was gathered.*

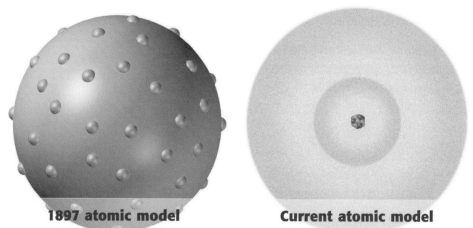

1897 atomic model   **Current atomic model**

Scientific Laws

What happens when a model correctly predicts the results of many different experiments? A scientific law can be constructed. In science, a **law** is a summary of many experimental results and observations. A law tells you how things work. Laws are not the same as theories. Laws tell you only what happens, not why it happens. Look at **Figure 5.** A chemical change took place when the flask was turned over. A light blue solid and a dark blue solution formed. Notice that the mass did not change, which demonstrates the *law of conservation of mass*. This law says that during a chemical change, the total mass of the materials formed is the same as the total mass of the starting materials. However, the law doesn't explain why. It tells you only what will happen during every chemical change.

Figure 5 *The total mass before the chemical change is always the same as the total mass after the change.*

SECTION Review

Summary

- A model uses familiar things to describe unfamiliar things.
- Physical, mathematical, and conceptual models are commonly used in science.
- A scientific theory is an explanation for many hypotheses and observations.
- A scientific law summarizes experimental results and observations. It describes what happens but not why.

Using Key Terms

1. In your own words, write a definition for the term *model*.

Understanding Key Ideas

2. Which kind of model would you use to represent a human heart?
   a. a mathematical model
   b. a physical model
   c. a conceptual model
   d. a natural model

3. Explain the difference between a theory and a law.

Critical Thinking

4. **Analyzing Methods** Both a globe and a flat world map can model features of Earth. Give an example of when you would use each of these models.

5. **Applying Concepts** Identify two limitations of physical models.

Math Skills

6. For a science fair, you want to make a model of the moon orbiting Earth by using two different balls. The diameter of the ball that will represent Earth will be about 62 cm. You want your model to be to scale. If the moon is about 4 times smaller than Earth, what should the diameter of the ball that represents the moon be?

SCiLINKS® **NSTA**
Developed and maintained by the
National Science Teachers Association

For a variety of links related to this chapter, go to www.scilinks.org

Topic: Using Models
SciLinks code: HSM1588

Tools, Measurement, and Safety

READING WARM-UP

Objectives

● Identify tools used to collect and analyze data.

● Explain the importance of the International System of Units.

● Identify the appropriate units to use for particular measurements.

● Identify safety symbols.

Terms to Learn

mass   density
volume   temperature

READING STRATEGY

Brainstorming The key idea of this section is scientific tools and measurements. Brainstorm what tools scientists use in their work and what the tools are used for.

Would you use a spoon to dig a hole to plant a tree? You wouldn't if you had a shovel!

To dig a hole, you need the correct tools. A *tool* is anything that helps you do a task. Scientists use many different tools to help them in their experiments.

Tools in Science

One way to collect data is to take measurements. To get the best measurements, you need the proper tools. Stopwatches, metersticks, and balances are some of the tools you can use to make measurements. Thermometers can be used to observe changes in temperature. Some of the uses for these tools are shown in **Figure 1**.

After you collect data, you need to analyze them. Calculators are handy tools to help you do calculations quickly. Or you might show your data in a graph or a figure. A computer that has the correct software can help you display your data. Of course, you can use a pencil and graph paper to graph your data.

✓ **Reading Check** Name two ways that scientists use tools. (*See the Appendix for answers to Reading Checks.*)

Making Measurements

Many years ago, different countries used different systems of measurement. In England, the standard for an inch used to be three grains of barley placed end to end. Other units were originally based on parts of the body, such as the foot.

Figure 1   **Measurement Tools**

You can use a stopwatch to measure time.

You can use a spring scale to measure force.

Table 1 Common SI Units

Length	meter (m)	
	kilometer (km)	1 km = 1,000 m
	decimeter (dm)	1 dm = 0.1 m
	centimeter (cm)	1 cm = 0.01 m
	millimeter (mm)	1 mm = 0.001 m
	micrometer (μm)	1 μm = 0.000 001 m
	nanometer (nm)	1 nm = 0.000 000 001 m
Volume	cubic meter (m^3)	
	cubic centimeter (cm^3)	1 cm^3 = 0.000 001 m^3
	liter (L)	1 L = 1 dm^3 = 0.001 m^3
	milliliter (mL)	1 mL = 0.001 L = 1 cm^3
Mass	kilogram (kg)	
	gram (g)	1 g = 0.001 kg
	milligram (mg)	1 mg = 0.000 001 kg
Temperature	Kelvin (K)	0°C = 273 K
	Celsius (°C)	100°C = 373 K

The International System of Units

In the late 1700s, the French Academy of Sciences set out to make a simple and reliable measurement system. Over the next 200 years, the metric system was formed. This system is now the International System of Units (SI). Because all SI units are expressed in multiples of 10, changing from one unit to another is easy. Prefixes are used to express SI units that are larger or smaller than basic units such as meter and gram. For example, *kilo-* means 1,000 times, and *milli-* indicates 1/1,000 times. The prefix used depends on the size of the object being measured. **Table 1** shows common SI units.

Length

To describe the length of an Olympic-sized swimming pool, a scientist would use meters (m). A *meter* is the basic SI unit of length. Other SI units of length are larger or smaller than the meter by multiples of 10. For example, if you divide 1 m into 1,000 parts, each part equals 1 millimeter (mm). So, 1 mm is one-thousandth of a meter.

Mass

Mass is the amount of matter in an object. The *kilogram* (kg) is the basic SI unit for mass. The kilogram is used to describe the mass of large objects. One kilogram equals 1,000 g. So, the gram is used to describe the mass of small objects. Masses of very large objects are expressed in metric tons. A metric ton equals 1,000 kg.

Units of Measure

Pick an object to use as a unit of measure. You can pick a pencil, your hand, or anything else. Find out how many units wide your desk is, and compare your measurement with those of your classmates. What were some of the units used? Now, choose two of the units that were used in your class, and make a conversion factor. For example, 1.5 pencils equal 1 board eraser.

mass a measure of the amount of matter in an object

Volume

Imagine that you need to move some lenses to a laser laboratory. How many lenses will fit into a crate? The answer depends on the volume of the crate and the volume of each lens. **Volume** is the amount of space that something occupies.

Liquid volume is expressed in *liters* (L). Liters are based on the meter. A cubic meter (1 m³) is equal to 1,000 L. So, 1,000 L will fit perfectly into a box that is 1 m on each side. A milliliter (mL) will fit perfectly into a box that is 1 cm on each side. So, 1 mL = 1 cm³. Graduated cylinders are used to measure the volume of liquids.

Volumes of solid objects are usually expressed in cubic meters (m³). Volumes of smaller objects can be expressed in cubic centimeters (cm³) or cubic millimeters (mm³). To find the volume of a crate—or any other rectangular shape—multiply the length by the width by the height.

Density

If you measure the mass and the volume of an object, you have the information you need to find the density of the object. **Density** is the amount of matter in a given volume. You cannot measure density directly. But after you measure the mass and the volume, you can calculate density by dividing the mass by the volume, as shown in the following equation:

$$D = \frac{m}{V}$$

Density is called a *derived quantity* because it is found by combining two basic quantities, mass and volume.

volume a measure of the size of an object or region in three-dimensional space

density the ratio of the mass of a substance to the volume of the substance

temperature a measure of how hot (or cold) something is; specifically, a measure of the average kinetic energy of the particles in an object

Figure 2 *Some common temperature measurements shown in degrees Fahrenheit and degrees Celsius*

Temperature

The **temperature** of a substance is a measurement of how hot (or cold) the substance is. Degrees Fahrenheit (°F) and degrees Celsius (°C) are used to describe temperature. However, the *kelvin* (K), the SI unit for temperature, is also used. Notice that the degree sign (°) is not used with the Kelvin scale. The thermometer in **Figure 2** shows how the Celsius and Fahrenheit scales compare.

✔ Reading Check What is the SI unit for temperature?

Safety Rules

Science is exciting and fun, but it can also be dangerous. Always follow your teacher's instructions. Don't take shortcuts, even when you think there is no danger. Read lab procedures carefully. Pay special attention to safety information and caution statements. **Figure 3** shows the safety symbols used in this book. Learn these symbols and their meanings by reading the safety information at the start of the book. If you are still not sure about what a safety symbol means, ask your teacher.

Figure 3   Safety Symbols

 Eye protection   Clothing protection   Hand safety Chemical safety Animal safety

 Heating safety   Electric safety   Sharp object Plant safety

SECTION Review

Summary

- Tools are used to make observations, take measurements, and analyze data.
- The International System of Units (SI) is the standard system of measurement.
- Length, volume, mass, and temperature are types of measurement.
- Density is the amount of matter in a given volume.
- Safety symbols are for your protection.

Using Key Terms

1. Use each of the following terms in a separate sentence: *volume, density,* and *mass.*

Understanding Key Ideas

2. Which SI unit would you use to express the height of your desk?
   a. kilogram   **c.** meter
   b. gram   **d.** liter

3. Explain the relationship between mass and density.

4. What is normal body temperature in degrees Fahrenheit and degrees Celsius?

Math Skills

5. A certain bacterial cell has a diameter of 0.50 μm. The tip of a pin is about 1,100 μm in diameter. How many of these bacterial cells would fit on the tip of the pin?

Critical Thinking

6. **Analyzing Ideas** What safety icons would you expect to see for a lab activity that asks you to pour acid into a beaker? Explain your answer.

7. **Applying Concepts** To find the area of a rectangle, multiply the length by the width. Why is area called a *derived quantity*?

SCiLINKS. NSTA
Developed and maintained by the
National Science Teachers Association

For a variety of links related to this chapter, go to www.scilinks.org

Topic: SI Units
SciLinks code: HSM1390

Skills Practice Lab

Measuring Liquid Volume

In this lab, you will use a graduated cylinder to measure and transfer precise amounts of liquids. Remember that, to accurately measure liquids in a graduated cylinder, you should first place the graduated cylinder flat on the lab table. Then, at eye level, read the volume of the liquid at the bottom of the meniscus, which is the curved surface of the liquid.

Procedure

1. Using the masking tape and marker, label the test tubes A, B, C, D, E, and F. Place them in the test-tube rack.

2. Make a data table as shown on the next page.

3. Using the graduated cylinder and the funnel, pour 14 mL of the red liquid into test tube A. (To do this, first measure out 10 mL of the liquid in the graduated cylinder, and pour it into the test tube. Then, measure an additional 4 mL of liquid in the graduated cylinder, and add this liquid to the test tube.)

4. Rinse the graduated cylinder and funnel with water each time you measure a different liquid.

5. Measure 13 mL of the yellow liquid, and pour it into test tube C.

6. Measure 13 mL of the blue liquid, and pour it into test tube E. Record the initial color and the volume of the liquid in each test tube.

OBJECTIVES

Measure accurately different volumes of liquids with a graduated cylinder.

Transfer exact amounts of liquids from a graduated cylinder to a test tube.

MATERIALS

- beakers, filled with colored liquid (3)
- funnel, small
- graduated cylinder, 10 mL
- marker
- tape, masking
- test-tube rack
- test tubes, large (6)

SAFETY

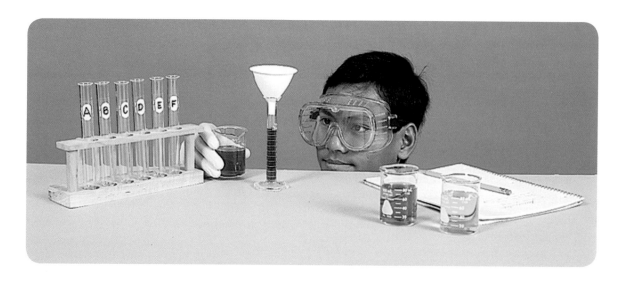

Data Table				
Test tube	Initial color	Initial volume	Final color	Final volume
A				
B				
C				
D				
E				
F				

DO NOT WRITE IN BOOK

7 Transfer 4 mL of liquid from test tube C into test tube D. Transfer 7 mL of liquid from test tube E into test tube D.

8 Measure 4 mL of blue liquid from the beaker, and pour it into test tube F. Measure 7 mL of red liquid from the beaker, and pour it into test tube F.

9 Transfer 8 mL of liquid from test tube A into test tube B. Transfer 3 mL of liquid from test tube C into test tube B.

Analyze the Results

1 Analyzing Data Record your final color observations in your data table.

2 Examining Data What is the final volume of all of the liquids? Use the graduated cylinder to measure the volume of liquid in each test tube. Record the volumes in your data table.

3 Organizing Data Record your final color observations and final volumes in a table of class data prepared by your teacher.

Draw Conclusions

4 Interpreting Information Did all of your classmates report the same colors? Form a hypothesis that could explain why the colors were the same or different after the liquids were combined.

5 Evaluating Methods Why should you not fill the graduated cylinder to the top?

Chapter Review

USING KEY TERMS

1 In your own words, write a definition for each of the following terms: *meter, temperature,* and *density*.

For each pair of terms, explain how the meanings of the terms differ.

2 *science* and *scientific methods*

3 *observation* and *hypothesis*

4 *theory* and *law*

5 *model* and *theory*

6 *volume* and *mass*

UNDERSTANDING KEY IDEAS

Multiple Choice

7 Physical science is
- **a.** the study of matter and energy.
- **b.** the study of physics and chemistry.
- **c.** important in most sciences.
- **d.** All of the above

8 The statement "Sheila has a stain on her shirt" is an example of a(n)
- **a.** law.
- **b.** hypothesis.
- **c.** observation.
- **d.** prediction.

9 A hypothesis
- **a.** may or may not be testable.
- **b.** is supported by evidence.
- **c.** is a possible answer to a question.
- **d.** All of the above

10 A variable
- **a.** is found in an uncontrolled experiment.
- **b.** is the factor that changes in an experiment.
- **c.** cannot change.
- **d.** is rarely included in experiments.

11 Organizing data into a graph is an example of
- **a.** collecting data.
- **b.** forming a hypothesis.
- **c.** asking a question.
- **d.** analyzing data.

12 How many milliliters are in 3.5 kL?
- **a.** 0.0035
- **c.** 35,000
- **b.** 3,500
- **d.** 3,500,000

13 A map of Seattle is an example of a
- **a.** physical model.
- **b.** mathematical model.
- **c.** conceptual model.
- **d.** All of the above

14 Ten meters is equal to
- **a.** 100 cm.
- **c.** 100,000 mm.
- **b.** 1,000 cm.
- **d.** 1,000 μm.

Short Answer

15 Describe three kinds of models used in science. Give an example and explain one limitation of each model.

16 Name two SI units that can be used to describe the volume of an object and two SI units that can be used to describe the mass of an object.

17 What are the steps used in scientific methods?

18 If a hypothesis is not testable, is the hypothesis wrong? Explain.

Math Skills

19 The cereal box on the right has a mass of 340 g. Its dimensions are 27 cm × 19 cm × 6 cm. What is the volume of the box? What is its density?

CRITICAL THINKING

20 Concept Mapping Use the following terms to create a concept map: *science, scientific methods, hypothesis, problems, questions, experiments,* and *observations*.

21 Applying Concepts A tailor is someone who makes or alters items of clothing. Why might a standard system of measurement be helpful to a tailor?

22 Analyzing Ideas Imagine that you are conducting an experiment. You are testing the effects of the height of a ramp on the speed at which a toy car goes down the ramp. What is the variable in this experiment? What factors must be controlled?

23 Evaluating Assumptions Suppose a classmate says, "I don't need to study science because I'm not going to be a scientist, and scientists are the only people who use science." How would you respond? In your answer, give several examples of careers that use physical science.

24 Making Inferences You build a model boat that you predict will float. However, your tests show that the boat sinks. What conclusion would you draw? Suggest some logical next steps.

INTERPRETING GRAPHICS

Use the picture below to answer the questions that follow.

25 How similar is this model to a real object?

26 What are some of the limitations of this model?

27 How might this model be useful?

Standardized Test Preparation

Read each of the passages below. Then, answer the questions that follow each passage.

Passage 1 The white light we see every day is actually composed of all of the colors of the visible spectrum. A laser emits a very small portion of this spectrum, so there can be blue lasers, red lasers, and so on. High-voltage sources called laser "pumps" cause laser materials to <u>emit</u> certain wavelengths of light depending on the material used. A laser material, such as a helium-neon (HeNe) gas mixture, emits radiation (light) as a result of electrons in high energy levels moving to lower energy levels. This process gives lasers their name: **l**ight **a**mplification of the **s**timulated **e**mission of **r**adiation.

1. Why are there blue lasers and red lasers?
   A White light is composed of all of the colors of the visible spectrum.
   B A laser emits a small portion of the visible spectrum.
   C A laser material emits radiation.
   D High-voltage sources are called laser "pumps."

2. In this passage, what is the meaning of the word *emit*?
   F to brighten
   G to compose
   H to change
   I to give off

3. Why does a laser produce radiation?
   A Only a small amount of light is used.
   B A laser is a high-voltage pump.
   C Light is made up of all of the colors in the visible spectrum.
   D Electrons in atoms change energy levels.

Passage 2 Researchers have created a new <u>class</u> of molecules. These molecules are called texaphyrins because of their large size and the five-pointed starlike shape at their center. Texaphyrins are similar to molecules that already exist in most living things. But texaphyrins are different because of their shape and their large size. The shape and large size of the molecules let scientists attach other elements to the molecules. Depending on what element is attached, texaphyrins can be used to locate tumors in the body or to help in treatments for some kinds of cancer.

1. Which of the following statements is true about texaphyrins, according to the passage?
   A They were just recently discovered.
   B They have the same shape that most natural molecules do.
   C They are used to treat certain cancers.
   D They are extremely small molecules.

2. In this passage, what is the meaning of the word *class*?
   F room
   G standing
   H rank
   I group

3. What is the main advantage of texaphyrin in treating tumors?
   A the small size of texaphyrin
   B the star shape of texaphyrin
   C the ability to attach to other substances
   D the man-made nature of the molecule

The graph below shows the changes in temperature during a chemical reaction. Use the graph below to answer the questions that follow.

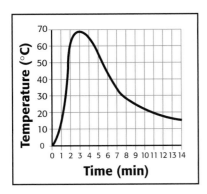

1. What was the highest temperature reached during the reaction?

   A 20°C

   B 40°C

   C 50°C

   D 70°C

2. During what period of time was the temperature increasing?

   F between 3 min and 14 min

   G between 0 min and 3 min

   H between 1 min and 13 min

   I between 0 min and 4 min

3. How many minutes did it take the temperature to increase from 10°C to 60°C?

   A less than 1 min

   B 1 min

   C 2 min

   D 3 min

4. About how many minutes passed from the time the highest temperature was reached until the time the temperature decreased to 20°C?

   F 7 min

   G 8 min

   H 11 min

   I 12 min

Read each question below, and choose the best answer.

1. What is the volume of a room that is 3.125 m high, 4.25 m wide, and 5.75 m long?

   A 13.1 m

   B 13.1 m³

   C 76.4 m

   D 76.4 m³

2. Yukiko has a storage box that measures 12 cm wide, 16.5 cm long, and 10 cm high. It has a mass of 850 g. What is the density of the box?

   F 1,980 cm³

   G 38.5 cm³

   H 2.3 g/cm³

   I .43 g/cm³

3. Remy traveled to Osaka, Japan, where the unit of currency is the yen. He spent 4,900 yen on train tickets. If the exchange rate was 113 yen to 1 U.S. dollar, approximately how much did the train tickets cost in U.S. dollars?

   A $25

   B $43

   C $49

   D $80

4. Lucia is measuring how fast bacteria grow in a Petri dish by measuring the area that the bacteria cover. On day 1, the bacteria cover 0.25 cm². On day 2, they cover 0.50 cm². On day 3, they cover 1.00 cm². What is the best prediction for the area covered on day 4?

   F 1.25 cm²

   G 1.50 cm²

   H 1.75 cm²

   I 2.00 cm²

Standardized Test Preparation

Science in Action

Science Fiction

"Inspiration" by Ben Bova

What if you were able to leap back and forth through time? Novelist H. G. Wells imagined such a possibility in his 1895 novelette *The Time Machine*. Most physicists said that time travel was against all the laws of physics. But what if Albert Einstein, then 16 and not a very good student, had met Wells and had an inspiration? Ben Bova's story "Inspiration" describes such a possibility. Young Einstein meets Wells and the great physicist of the time, Lord Kelvin. But was the meeting just a lucky coincidence or something else entirely? Escape to the *Holt Anthology of Science Fiction*, and read "Inspiration."

Social Studies ACTiViTY

Research the life of Albert Einstein from high school through college. Make a poster that describes some of his experiences during this time. Include information about how he matured as a student.

Weird Science

A Palace of Ice

An ice palace is just a fancy kind of igloo, but it takes a lot of ice and snow to make an ice palace. One ice palace was made from 27,215.5 metric tons of snow and 9,071.85 metric tons of ice! Making an ice palace takes time, patience, and temperatures below freezing. Sometimes, blocks of ice are cut with chain saws from a frozen river or lake and then transported in huge trucks. On location, the huge ice cubes are stacked on each other. Slush is used as mortar between the "bricks." The slush freezes and cements the blocks of ice together. Then, sculptors with chain saws, picks, and axes fashion elegant details in the ice.

Math ACTiViTY

One block of ice used to make the ice palace in the story above has a mass of 181.44 kg. How many blocks of ice were needed to make the ice palace if 9071.85 metric tons of ice was used?

Careers

Julie Williams-Byrd

Electronics Engineer Julie Williams-Byrd uses her knowledge of physics to develop better lasers. She started working with lasers when she was a graduate student at Hampton University in Virginia. Today, Williams-Byrd works as an electronics engineer in the Laser Systems Branch (LSB) of NASA. She designs and builds lasers that are used to study wind and ozone in the atmosphere. Williams-Byrd uses scientific models to predict the nature of different aspects of laser design. For example, laser models are used to predict output energy, wavelength, and efficiency of the laser system.

Her most challenging project has been building a laser transmitter that will be used to measure winds in the atmosphere. This system, called *Lidar,* is very much like radar except that it uses light waves instead of sound waves to bounce off objects. Although Williams-Byrd works with high-tech lasers, she points out that lasers are a part of daily life for many people. For example, lasers are used in scanners at many retail stores. Ophthalmologists use lasers to correct vision problems. Some metal workers use them to cut metal. And lasers are even used to create spectacular light shows!

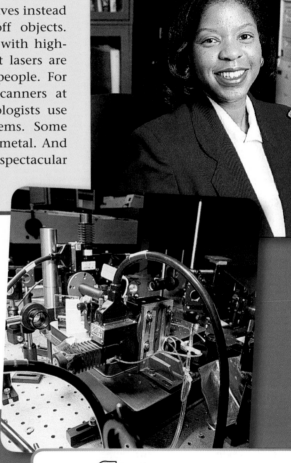

Language Arts ACTiViTY

WRITING SKILL Research lasers and how they can be used in everyday life. Then, write a one-page essay on how lasers have made life easier for people.

To learn more about these Science in Action topics, visit **go.hrw.com** and type in the keyword **HP5WPSF**.

Current Science

Check out Current Science® articles related to this chapter by visiting **go.hrw.com**. Just type in the keyword **HP5CS01**.

Skills Practice Lab

Graphing Data

When performing an experiment, you usually need to collect data. To understand the data, you can often organize them into a graph. Graphs can show trends and patterns that you might not notice in a table or list. In this exercise, you will practice collecting data and organizing the data into a graph.

MATERIALS

- beaker, 400 mL
- clock (or watch) with a second hand
- gloves, heat-resistant
- hot plate
- ice
- paper, graph
- thermometer, Celsius, with a clip
- water, 200 mL

SAFETY

Procedure

1. Pour 200 mL of water into a 400 mL beaker. Add ice to the beaker until the waterline is at the 400 mL mark.

2. Place a Celsius thermometer into the beaker. Use a thermometer clip to prevent the thermometer from touching the bottom of the beaker. Record the temperature of the ice water.

3. Place the beaker and thermometer on a hot plate. Turn the hot plate on medium heat, and record the temperature every minute until the water temperature reaches 100°C.

4. Using heat-resistant gloves, remove the beaker from the hot plate. Continue to record the temperature of the water each minute for 10 more minutes. **Caution:** Don't forget to turn off the hot plate.

5. On a piece of graph paper, create a graph similar to the one below. Label the horizontal axis (the x-axis) "Time (min)," and mark the axis in increments of 1 min as shown. Label the vertical axis (the y-axis) "Temperature (°C)," and mark the axis in increments of 10° as shown.

6. Find the 1 min mark on the x-axis, and move up the graph to the temperature you recorded at 1 min. Place a dot on the graph at that point. Plot each temperature in the same way. When you have plotted all of your data, connect the dots with a smooth line.

Analyze the Results

1. Examine your graph. Do you think the water heated faster than it cooled? Explain.

2. Estimate what the temperature of the water was 2.5 min after you placed the beaker on the hot plate. Explain how you can make a good estimate of temperature between those you recorded.

Draw Conclusions

3. Explain how a graph may give more information than the same data in a table.

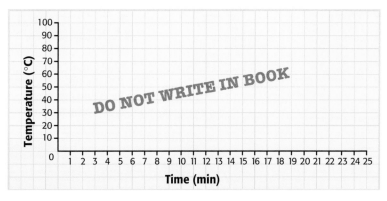

Model-Making Lab

A Window to a Hidden World

Have you ever noticed that objects underwater appear closer than they really are? The reason is that light waves change speed when they travel from air into water. Anton van Leeuwenhoek, a pioneer of microscopy in the late 17th century, used a drop of water to magnify objects. That drop of water brought a hidden world closer into view. How did Leeuwenhoek's microscope work? In this investigation, you will build a model of it to find out.

MATERIALS

- eyedropper
- hole punch
- newspaper
- plastic wrap, clear
- poster board, 3 cm × 10 cm
- tape, transparent
- water

Procedure

1. Punch a hole in the center of the poster board with a hole punch, as shown in (a) at right.

2. Tape a small piece of clear plastic wrap over the hole, as shown in (b) at right. Be sure the plastic wrap is large enough so that the tape you use to secure it does not cover the hole.

3. Use an eyedropper to put one drop of water over the hole. Check to be sure your drop of water is dome-shaped (convex), as shown in (c) at right.

4. Hold the microscope close to your eye and look through the drop. Be careful not to disturb the water drop.

5. Hold the microscope over a piece of newspaper, and observe the image.

Analyze the Results

1. Describe and draw the image you see. Is the image larger than or the same size as it is without the microscope? Is the image clear or blurred? Is the shape of the image distorted?

Draw Conclusions

2. How do you think your model could be improved?

Applying Your Data

Robert Hooke and Zacharias Janssen contributed much to the field of microscopy. Research one of them, and write a paragraph about his contributions.

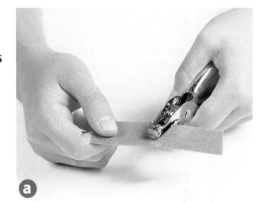

a

b

c

Skills Practice Lab

Exploring the Unseen

Your teacher will give you a box in which a special divider has been created. Your task is to describe this divider as precisely as possible—without opening the box! Your only aid is a marble that is also inside the box. This task will allow you to demonstrate your understanding of the scientific method. Good luck!

MATERIALS

• mystery box, sealed

Ask a Question

1 Record the question that you are trying to answer by doing this experiment. (Hint: Read the introductory paragraph again if you are not sure what your task is.)

Form a Hypothesis

2 Before you begin the experiment, think about what's required. Do you think you will be able to easily determine the shape of the divider? Can you determine its texture or color? Write a hypothesis that states how much you think you will be able to determine about the divider during the experiment. (Remember that you can't open the box!)

Test the Hypothesis

3 Using all the methods you can think of (except opening the box), test your hypothesis. Make careful notes about your testing and observations.

Analyze the Results

1 What characteristics of the divider were you able to identify? Draw or write your best description of the interior of the box.

2 Do your observations support your hypothesis? Explain. If your results do not support your hypothesis, write a new hypothesis, and test it.

3 With your teacher's permission, open the box, and look inside. Record your observations.

Draw Conclusions

4 Write a paragraph summarizing your experiment. Be sure to include what methods you used, whether your results supported your hypothesis, and how you could improve your methods.

Model-Making Lab

Off to the Races!

Scientists often use models—representations of objects or systems. Physical models, such as a model airplane, are generally a different size than the objects they represent. In this lab, you will build a model car, test its design, and then try to improve the design.

MATERIALS

- board
- clothes-hanger wire, 16 cm
- eraser, pink rubber, or small wood block
- glue
- paper, typing (2 sheets)
- pliers (or wire cutters)
- ruler, metric
- stopwatch
- textbooks

SAFETY

Procedure

1. Using the materials listed, design and build a car that will carry the load (the eraser or block of wood) down the ramp as quickly as possible. Your car must be no wider than 8 cm, it must have room to carry the load, and it must roll.

2. As you test your design, do not be afraid to rebuild or re-design your car. Improving your methods is an important part of scientific progress.

3. When you have a design that works well, measure the time required for your car to roll down the ramp. Record this time. Test your car with this design several times for accuracy.

4. Try to improve your model. Find one thing that you can change to make your model car roll faster down the ramp. Write a description of the change.

5. Test your model again as you did in step 3 and make additional improvements if needed.

Analyze the Results

1. Why is it important to have room in the model car for the eraser or wood block? (Hint: Think about the function of a real car.)

2. Before you built the model car, you created a design for it. Do you think this design is also a model? Explain.

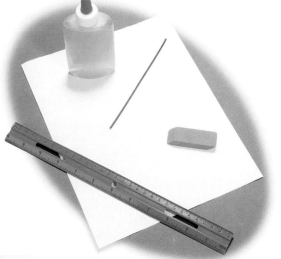

3. Based on your observations in this lab, list three reasons why it is helpful for automobile designers to build and test small model cars rather than immediately build a full-size car.

Draw Conclusions

4. In this lab, you built a model that was smaller than the object it represented. Some models are larger than the objects they represent. List three examples of larger models that are used to represent objects. Why is it helpful to use a larger model in these cases?

Skills Practice Lab

Coin Operated

All pennies are exactly the same, right? Probably not! After all, each penny was made in a certain year at a specific mint, and each has traveled a unique path to reach your classroom. But all pennies are similar. In this lab, you will investigate differences and similarities among a group of pennies.

MATERIALS

- balance, metric
- graduated cylinder, 100 mL
- paper, notebook (10 sheets)
- paper towels
- pennies (10)
- water

SAFETY

Procedure

1 Write the numbers 1 through 10 on a page, and place a penny next to each number.

2 Use the metric balance to find the mass of each penny to the nearest 0.1 g. Record each measurement next to the number of that penny.

3 On a table that your teacher will provide, make a mark in the correct column of the table for each penny you measured.

4 Separate your pennies into piles, based on the class data. Place each pile on its own sheet of paper.

5 Measure and record the mass of each pile. Write the mass on the paper you are using to identify the pile.

6 Fill a graduated cylinder halfway with water. Carefully measure the volume in the cylinder, and record it.

7 Carefully place the pennies from one pile into the graduated cylinder. Measure and record the new volume.

8 Carefully pour out the water into the sink, and remove the pennies from the graduated cylinder. With a paper towel, dry off the pile of pennies.

9 Repeat steps 6 through 8 for each pile of pennies.

Analyze the Results

1 Determine the volume of the displaced water by subtracting the initial volume from the final volume. This amount is equal to the volume of the pennies. Record the volume of each pile of pennies.

2 Calculate the density of each pile. To make this calculation, divide the total mass of the pennies by the volume of the pennies. Record the density.

3 What differences, if any, did you note in the mass, volume, and density of the pennies?

Draw Conclusions

4 If you noted differences, what do you think might be the cause of these differences?

5 How is it possible for the pennies to have different densities?

6 What clues might allow you to separate the pennies into the same groups without experimentation? Explain.

Contents

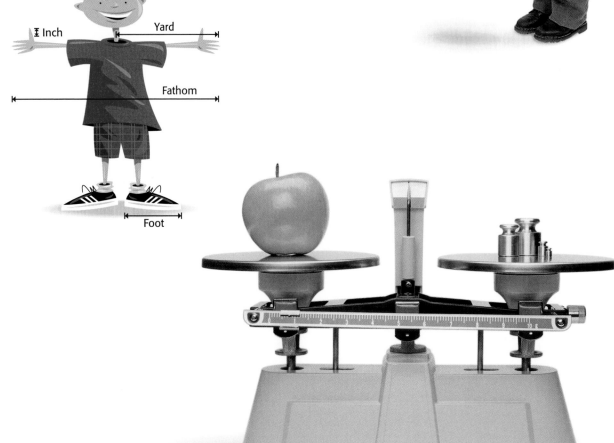

Appendix

✔ Reading Check Answers

Chapter 1 The World of Life Science

Section 1
Page 4: the study of living things

Page 7: Sample answer: ocean pollution that harms mammals, birds, and fish.

Section 2
Page 8: a series of steps used by scientists to solve problems

Page 10: the possibility that an experiment can be designed to test the hypothesis

Page 12: only one

Page 14: because the scientist has learned something

Section 3
Page 17: a mathematical model

Page 18: to explain a broad range of observations, facts, and tested hypotheses, to predict what might happen, and to organize scientific thinking

Section 4
Page 21: SEMs produce three-dimensional images, and TEMs produce flat images.

Page 23: square units, such as square meters (m^2) and square centimeters (cm^2)

Page 24: how hot or cold it is or how much energy it has

Chapter 2 The World of Earth Science

Section 1
Page 37: Four areas of oceanography are physical oceanography, biological oceanography, geological oceanography, and chemical oceanography.

Page 39: Astronomers study stars, asteroids, planets, and everything else in space.

Page 41: Cartographers make maps.

Section 2
Page 42: Scientists begin to learn about things by asking questions.

Page 45: Scientists create graphs and tables to organize and summarize their data.

Page 46: It is important for the scientific community to review new evidence so that scientists can evaluate and question the evidence for accuracy.

Section 3
Page 49: The big bang theory is an explanation of the creation of the universe.

Page 51: A climate model is complicated because there are so many variables that affect climate.

Section 4
Page 52: The International System of Units was developed to create a standard measurement system.

Page 55: Before you start a science investigation, obtain your teacher's permission and read the lab procedures carefully.

Chapter 3 The World of Physical Science

Section 1

Page 66: The first step in gathering knowledge is asking a question.

Page 68: how matter interacts with other matter, what the structure and properties of matter are, and how substances change

Page 70: A geochemist studies the chemistry of rocks, minerals, and soil.

Section 2

Page 73: Asking questions helps focus the purpose of the investigation.

Page 74: Boat efficiency is important because it saves resources, such as fuel.

Page 77: Information is easier to see and understand when it is organized into charts and graphs.

Page 78: The results of an investigation can be communicated by writing a scientific paper, making a presentation, or creating a Web site.

Section 3

Page 81: One possible limitation of a mathematical model is that complex models may have unknown variables. If the unknown variables change, a mathematical model could fail.

Page 82: A theory can explain a hypothesis or an observation.

Section 4

Page 84: Scientists use tools to make measurements and analyze data.

Page 86: The SI unit for temperature is the kelvin (K).

Study Skills

FoldNote Instructions

Have you ever tried to study for a test or quiz but didn't know where to start? Or have you read a chapter and found that you can remember only a few ideas? Well, FoldNotes are a fun and exciting way to help you learn and remember the ideas you encounter as you learn science!

FoldNotes are tools that you can use to organize concepts. By focusing on a few main concepts, FoldNotes help you learn and remember how the concepts fit together. They can help you see the "big picture." Below you will find instructions for building 10 different FoldNotes.

Pyramid

1. Place a sheet of paper in front of you. Fold the lower left-hand corner of the paper diagonally to the opposite edge of the paper.

2. Cut off the tab of paper created by the fold (at the top).

3. Open the paper so that it is a square. Fold the lower right-hand corner of the paper diagonally to the opposite corner to form a triangle.

4. Open the paper. The creases of the two folds will have created an X.

5. Using scissors, cut along one of the creases. Start from any corner, and stop at the center point to create two flaps. Use tape or glue to attach one of the flaps on top of the other flap.

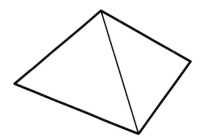

Double Door

1. Fold a sheet of paper in half from the top to the bottom. Then, unfold the paper.

2. Fold the top and bottom edges of the paper to the crease.

Booklet

1. Fold a sheet of paper in half from left to right. Then, unfold the paper.

2. Fold the sheet of paper in half again from the top to the bottom. Then, unfold the paper.

3. Refold the sheet of paper in half from left to right.

4. Fold the top and bottom edges to the center crease.

5. Completely unfold the paper.

6. Refold the paper from top to bottom.

7. Using scissors, cut a slit along the center crease of the sheet from the folded edge to the creases made in step 4. Do not cut the entire sheet in half.

8. Fold the sheet of paper in half from left to right. While holding the bottom and top edges of the paper, push the bottom and top edges together so that the center collapses at the center slit. Fold the four flaps to form a four-page book.

Layered Book

1. Lay one sheet of paper on top of another sheet. Slide the top sheet up so that 2 cm of the bottom sheet is showing.

2. Hold the two sheets together, fold down the top of the two sheets so that you see four 2 cm tabs along the bottom.

3. Using a stapler, staple the top of the FoldNote.

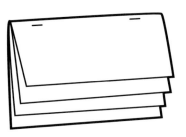

Key-Term Fold

1. Fold a sheet of lined notebook paper in half from left to right.

2. Using scissors, cut along every third line from the right edge of the paper to the center fold to make tabs.

Four-Corner Fold

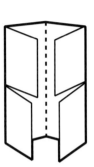

1. Fold a sheet of paper in half from left to right. Then, unfold the paper.

2. Fold each side of the paper to the crease in the center of the paper.

3. Fold the paper in half from the top to the bottom. Then, unfold the paper.

4. Using scissors, cut the top flap creases made in step 3 to form four flaps.

Three-Panel Flip Chart

1. Fold a piece of paper in half from the top to the bottom.

2. Fold the paper in thirds from side to side. Then, unfold the paper so that you can see the three sections.

3. From the top of the paper, cut along each of the vertical fold lines to the fold in the middle of the paper. You will now have three flaps.

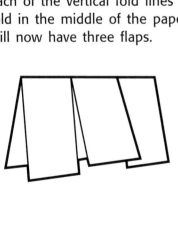

Appendix

Table Fold

1. Fold a piece of paper in half from the top to the bottom. Then, fold the paper in half again.

2. Fold the paper in thirds from side to side.

3. Unfold the paper completely. Carefully trace the fold lines by using a pen or pencil.

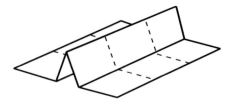

Two-Panel Flip Chart

1. Fold a piece of paper in half from the top to the bottom.

2. Fold the paper in half from side to side. Then, unfold the paper so that you can see the two sections.

3. From the top of the paper, cut along the vertical fold line to the fold in the middle of the paper. You will now have two flaps.

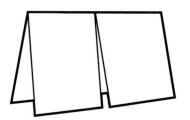

Tri-Fold

1. Fold a piece a paper in thirds from the top to the bottom.

2. Unfold the paper so that you can see the three sections. Then, turn the paper sideways so that the three sections form vertical columns.

3. Trace the fold lines by using a pen or pencil. Label the columns "Know," "Want," and "Learn."

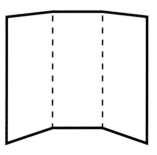

Graphic Organizer Instructions

Have you ever wished that you could "draw out" the many concepts you learn in your science class? Sometimes, being able to *see* how concepts are related really helps you remember what you've learned. Graphic Organizers do just that! They give you a way to draw or map out concepts.

All you need to make a Graphic Organizer is a piece of paper and a pencil. Below you will find instructions for four different Graphic Organizers designed to help you organize the concepts you'll learn in this book.

Spider Map

1. Draw a diagram like the one shown. In the circle, write the main topic.

2. From the circle, draw legs to represent different categories of the main topic. You can have as many categories as you want.

3. From the category legs, draw horizontal lines. As you read the chapter, write details about each category on the horizontal lines.

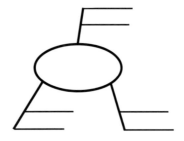

Comparison Table

1. Draw a chart like the one shown. Your chart can have as many columns and rows as you want.

2. In the top row, write the topics that you want to compare.

3. In the left column, write characteristics of the topics that you want to compare. As you read the chapter, fill in the characteristics for each topic in the appropriate boxes.

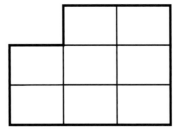

Chain-of-Events-Chart

1. Draw a box. In the box, write the first step of a process or the first event of a timeline.

2. Under the box, draw another box, and use an arrow to connect the two boxes. In the second box, write the next step of the process or the next event in the timeline.

3. Continue adding boxes until the process or timeline is finished.

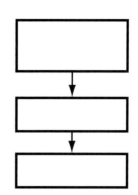

Concept Map

1. Draw a circle in the center of a piece of paper. Write the main idea of the chapter in the center of the circle.

2. From the circle, draw other circles. In those circles, write characteristics of the main idea. Draw arrows from the center circle to the circles that contain the characteristics.

3. From each circle that contains a characteristic, draw other circles. In those circles, write specific details about the characteristic. Draw arrows from each circle that contains a characteristic to the circles that contain specific details. You may draw as many circles as you want.

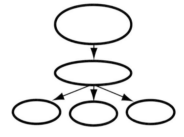

SI Measurement

The International System of Units, or SI, is the standard system of measurement used by many scientists. Using the same standards of measurement makes it easier for scientists to communicate with one another.

SI works by combining prefixes and base units. Each base unit can be used with different prefixes to define smaller and larger quantities. The table below lists common SI prefixes.

SI Prefixes

Prefix	Symbol	Factor	Example
kilo-	k	1,000	kilogram, 1 kg = 1,000 g
hecto-	h	100	hectoliter, 1 hL = 100 L
deka-	da	10	dekameter, 1 dam = 10 m
		1	meter, liter, gram
deci-	d	0.1	decigram, 1 dg = 0.1 g
centi-	c	0.01	centimeter, 1 cm = 0.01 m
milli-	m	0.001	milliliter, 1 mL = 0.001 L
micro-	μ	0.000 001	micrometer, 1 μm = 0.000 001 m

SI Conversion Table

SI units	From SI to English	From English to SI
Length		
kilometer (km) = 1,000 m	1 km = 0.621 mi	1 mi = 1.609 km
meter (m) = 100 cm	1 m = 3.281 ft	1 ft = 0.305 m
centimeter (cm) = 0.01 m	1 cm = 0.394 in.	1 in. = 2.540 cm
millimeter (mm) = 0.001 m	1 mm = 0.039 in.	
micrometer (μm) = 0.000 001 m		
nanometer (nm) = 0.000 000 001 m		
Area		
square kilometer (km^2) = 100 hectares	1 km^2 = 0.386 mi^2	1 mi^2 = 2.590 km^2
hectare (ha) = 10,000 m^2	1 ha = 2.471 acres	1 acre = 0.405 ha
square meter (m^2) = 10,000 cm^2	1 m^2 = 10.764 ft^2	1 ft^2 = 0.093 m^2
square centimeter (cm^2) = 100 mm^2	1 cm^2 = 0.155 $in.^2$	1 $in.^2$ = 6.452 cm^2
Volume		
liter (L) = 1,000 mL = 1 dm^3	1 L = 1.057 fl qt	1 fl qt = 0.946 L
milliliter (mL) = 0.001 L = 1 cm^3	1 mL = 0.034 fl oz	1 fl oz = 29.574 mL
microliter (μL) = 0.000 001 L		
Mass		
kilogram (kg) = 1,000 g	1 kg = 2.205 lb	1 lb = 0.454 kg
gram (g) = 1,000 mg	1 g = 0.035 oz	1 oz = 28.350 g
milligram (mg) = 0.001 g		
microgram (μg) = 0.000 001 g		

Appendix

Temperature Scales

Temperature can be expressed by using three different scales: Fahrenheit, Celsius, and Kelvin. The SI unit for temperature is the kelvin (K).

Although 0 K is much colder than 0°C, a change of 1 K is equal to a change of 1°C.

Three Temperature Scales

	Fahrenheit	Celsius	Kelvin
Water boils	212°	100°	373
Body temperature	98.6°	37°	310
Room temperature	68°	20°	293
Water freezes	32°	0°	273

Temperature Conversions Table

To convert	Use this equation:	Example
Celsius to Fahrenheit °C → °F	$°F = \left(\dfrac{9}{5} \times °C\right) + 32$	Convert 45°C to °F. $°F = \left(\dfrac{9}{5} \times 45°C\right) + 32 = 113°F$
Fahrenheit to Celsius °F → °C	$°C = \dfrac{5}{9} \times (°F - 32)$	Convert 68°F to °C. $°C = \dfrac{5}{9} \times (68°F - 32) = 20°C$
Celsius to Kelvin °C → K	$K = °C + 273$	Convert 45°C to K. $K = 45°C + 273 = 318\ K$
Kelvin to Celsius K → °C	$°C = K - 273$	Convert 32 K to °C. $°C = 32K - 273 = -241°C$

Scientific Methods

The ways in which scientists answer questions and solve problems are called **scientific methods.** The same steps are often used by scientists as they look for answers. However, there is more than one way to use these steps. Scientists may use all of the steps or just some of the steps during an investigation. They may even repeat some of the steps. The goal of using scientific methods is to come up with reliable answers and solutions.

Six Steps of Scientific Methods

1 Ask a Question

Good questions come from careful **observations.** You make observations by using your senses to gather information. Sometimes, you may use instruments, such as microscopes and telescopes, to extend the range of your senses. As you observe the natural world, you will discover that you have many more questions than answers. These questions drive investigations.

Questions beginning with *what, why, how,* and *when* are important in focusing an investigation. Here is an example of a question that could lead to an investigation.

Question: How does acid rain affect plant growth?

2 Form a Hypothesis

After you ask a question, you need to form a **hypothesis.** A hypothesis is a clear statement of what you expect the answer to your question to be. Your hypothesis will represent your best "educated guess" based on what you have observed and what you already know. A good hypothesis is testable. Otherwise, the investigation can go no further. Here is a hypothesis based on the question, "How does acid rain affect plant growth?"

Hypothesis: Acid rain slows plant growth.

The hypothesis can lead to predictions. A prediction is what you think the outcome of your experiment or data collection will be. Predictions are usually stated in an if-then format. Here is a sample prediction for the hypothesis that acid rain slows plant growth.

Prediction: If a plant is watered with only acid rain (which has a pH of 4), then the plant will grow at half its normal rate.

3 Test the Hypothesis

After you have formed a hypothesis and made a prediction, your hypothesis should be tested. One way to test a hypothesis is with a controlled experiment. A **controlled experiment** tests only one factor at a time. In an experiment to test the effect of acid rain on plant growth, the **control group** would be watered with normal rain water. The **experimental group** would be watered with acid rain. All of the plants should receive the same amount of sunlight and water each day. The air temperature should be the same for all groups. However, the acidity of the water will be a variable. In fact, any factor that is different from one group to another is a **variable.** If your hypothesis is correct, then the acidity of the water and plant growth are *dependant variables.* The amount a plant grows is dependent on the acidity of the water. However, the amount of water each plant receives and the amount of sunlight each plant receives are *independent variables.* Either of these factors could change without affecting the other factor.

Sometimes, the nature of an investigation makes a controlled experiment impossible. For example, the Earth's core is surrounded by thousands of meters of rock. Under such circumstances, a hypothesis may be tested by making detailed observations.

4 Analyze the Results

After you have completed your experiments, made your observations, and collected your data, you must analyze all the information you have gathered. Tables and graphs are often used in this step to organize the data.

5 Draw Conclusions

After analyzing your data, you can determine if your results support your hypothesis. If your hypothesis is supported, you (or others) might want to repeat the observations or experiments to verify your results. If your hypothesis is not supported by the data, you may have to check your procedure for errors. You may even have to reject your hypothesis and make a new one. If you cannot draw a conclusion from your results, you may have to try the investigation again or carry out further observations or experiments.

6 Communicate Results

After any scientific investigation, you should report your results. By preparing a written or oral report, you let others know what you have learned. They may repeat your investigation to see if they get the same results. Your report may even lead to another question and then to another investigation.

Scientific Methods in Action

Scientific methods contain loops in which several steps may be repeated over and over again. In some cases, certain steps are unnecessary. Thus, there is not a "straight line" of steps. For example, sometimes scientists find that testing one hypothesis raises new questions and new hypotheses to be tested. And sometimes, testing the hypothesis leads directly to a conclusion. Furthermore, the steps in scientific methods are not always used in the same order. Follow the steps in the diagram, and see how many different directions scientific methods can take you.

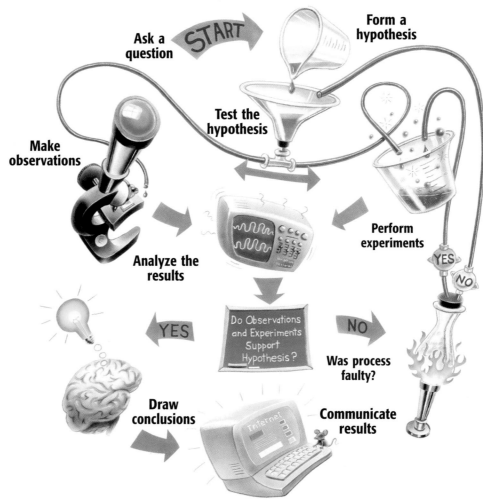

Measuring Skills

Using a Graduated Cylinder

When using a graduated cylinder to measure volume, keep the following procedures in mind:

1. Place the cylinder on a flat, level surface before measuring liquid.

2. Move your head so that your eye is level with the surface of the liquid.

3. Read the mark closest to the liquid level. On glass graduated cylinders, read the mark closest to the center of the curve in the liquid's surface.

Using a Meterstick or Metric Ruler

When using a meterstick or metric ruler to measure length, keep the following procedures in mind:

1. Place the ruler firmly against the object that you are measuring.

2. Align one edge of the object exactly with the 0 end of the ruler.

3. Look at the other edge of the object to see which of the marks on the ruler is closest to that edge. (Note: Each small slash between the centimeters represents a millimeter, which is one-tenth of a centimeter.)

Using a Triple-Beam Balance

When using a triple-beam balance to measure mass, keep the following procedures in mind:

1. Make sure the balance is on a level surface.

2. Place all of the countermasses at 0. Adjust the balancing knob until the pointer rests at 0.

3. Place the object you wish to measure on the pan. **Caution:** Do not place hot objects or chemicals directly on the balance pan.

4. Move the largest countermass along the beam to the right until it is at the last notch that does not tip the balance. Follow the same procedure with the next-largest countermass. Then, move the smallest countermass until the pointer rests at 0.

5. Add the readings from the three beams together to determine the mass of the object.

6. When determining the mass of crystals or powders, first find the mass of a piece of filter paper. Then, add the crystals or powder to the paper, and remeasure. The actual mass of the crystals or powder is the total mass minus the mass of the paper. When finding the mass of liquids, first find the mass of the empty container. Then, find the combined mass of the liquid and container. The mass of the liquid is the total mass minus the mass of the container.

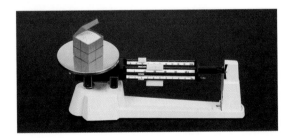

Making Charts and Graphs

Pie Charts

A pie chart shows how each group of data relates to all of the data. Each part of the circle forming the chart represents a category of the data. The entire circle represents all of the data. For example, a biologist studying a hardwood forest in Wisconsin found that there were five different types of trees. The data table at right summarizes the biologist's findings.

Wisconsin Hardwood Trees	
Type of tree	**Number found**
Oak	600
Maple	750
Beech	300
Birch	1,200
Hickory	150
Total	3,000

How to Make a Pie Chart

1 To make a pie chart of these data, first find the percentage of each type of tree. Divide the number of trees of each type by the total number of trees, and multiply by 100.

$$\frac{600 \text{ oak}}{3,000 \text{ trees}} \times 100 = 20\%$$

$$\frac{750 \text{ maple}}{3,000 \text{ trees}} \times 100 = 25\%$$

$$\frac{300 \text{ beech}}{3,000 \text{ trees}} \times 100 = 10\%$$

$$\frac{1,200 \text{ birch}}{3,000 \text{ trees}} \times 100 = 40\%$$

$$\frac{150 \text{ hickory}}{3,000 \text{ trees}} \times 100 = 5\%$$

2 Now, determine the size of the wedges that make up the pie chart. Multiply each percentage by 360°. Remember that a circle contains 360°.

$20\% \times 360° = 72°$ $25\% \times 360° = 90°$

$10\% \times 360° = 36°$ $40\% \times 360° = 144°$

$5\% \times 360° = 18°$

3 Check that the sum of the percentages is 100 and the sum of the degrees is 360.

$20\% + 25\% + 10\% + 40\% + 5\% = 100\%$

$72° + 90° + 36° + 144° + 18° = 360°$

4 Use a compass to draw a circle and mark the center of the circle.

5 Then, use a protractor to draw angles of 72°, 90°, 36°, 144°, and 18° in the circle.

6 Finally, label each part of the chart, and choose an appropriate title.

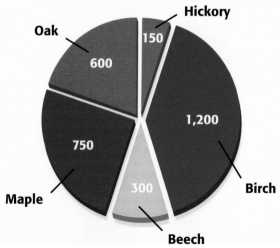

A Community of Wisconsin Hardwood Trees

Line Graphs

Line graphs are most often used to demonstrate continuous change. For example, Mr. Smith's students analyzed the population records for their hometown, Appleton, between 1900 and 2000. Examine the data at right.

Because the year and the population change, they are the *variables*. The population is determined by, or dependent on, the year. Therefore, the population is called the **dependent variable,** and the year is called the **independent variable.** Each set of data is called a **data pair.** To prepare a line graph, you must first organize data pairs into a table like the one at right.

Population of Appleton, 1900–2000	
Year	**Population**
1900	1,800
1920	2,500
1940	3,200
1960	3,900
1980	4,600
2000	5,300

How to Make a Line Graph

1 Place the independent variable along the horizontal (*x*) axis. Place the dependent variable along the vertical (*y*) axis.

2 Label the *x*-axis "Year" and the *y*-axis "Population." Look at your largest and smallest values for the population. For the *y*-axis, determine a scale that will provide enough space to show these values. You must use the same scale for the entire length of the axis. Next, find an appropriate scale for the *x*-axis.

3 Choose reasonable starting points for each axis.

4 Plot the data pairs as accurately as possible.

5 Choose a title that accurately represents the data.

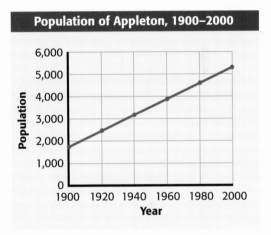

How to Determine Slope

Slope is the ratio of the change in the *y*-value to the change in the *x*-value, or "rise over run."

1 Choose two points on the line graph. For example, the population of Appleton in 2000 was 5,300 people. Therefore, you can define point *a* as (2000, 5,300). In 1900, the population was 1,800 people. You can define point *b* as (1900, 1,800).

2 Find the change in the *y*-value. (*y* at point *a*) − (*y* at point *b*) = 5,300 people − 1,800 people = 3,500 people

3 Find the change in the *x*-value. (*x* at point *a*) − (*x* at point *b*) = 2000 − 1900 = 100 years

4 Calculate the slope of the graph by dividing the change in *y* by the change in *x*.

$$slope = \frac{change\ in\ y}{change\ in\ x}$$

$$slope = \frac{3,500\ people}{100\ years}$$

$$slope = 35\ people\ per\ year$$

In this example, the population in Appleton increased by a fixed amount each year. The graph of these data is a straight line. Therefore, the relationship is **linear.** When the graph of a set of data is not a straight line, the relationship is **nonlinear.**

Using Algebra to Determine Slope

The equation in step 4 may also be arranged to be

$$y = kx$$

where y represents the change in the y-value, k represents the slope, and x represents the change in the x-value.

$$slope = \frac{change\ in\ y}{change\ in\ x}$$

$$k = \frac{y}{x}$$

$$k \times x = \frac{y \times x}{x}$$

$$kx = y$$

Bar Graphs

Bar graphs are used to demonstrate change that is not continuous. These graphs can be used to indicate trends when the data cover a long period of time. A meteorologist gathered the precipitation data shown here for Hartford, Connecticut, for April 1–15, 1996, and used a bar graph to represent the data.

Precipitation in Hartford, Connecticut April 1–15, 1996			
Date	Precipitation (cm)	Date	Precipitation (cm)
April 1	0.5	April 9	0.25
April 2	1.25	April 10	0.0
April 3	0.0	April 11	1.0
April 4	0.0	April 12	0.0
April 5	0.0	April 13	0.25
April 6	0.0	April 14	0.0
April 7	0.0	April 15	6.50
April 8	1.75		

How to Make a Bar Graph

1. Use an appropriate scale and a reasonable starting point for each axis.

2. Label the axes, and plot the data.

3. Choose a title that accurately represents the data.

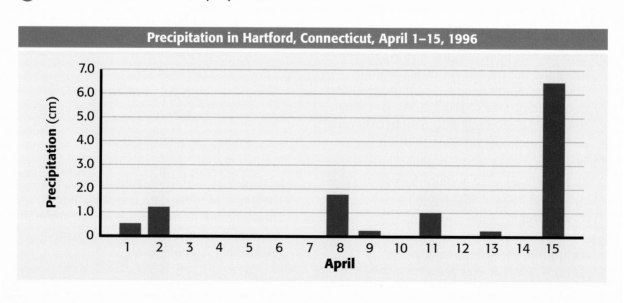

Math Refresher

Science requires an understanding of many math concepts. The following pages will help you review some important math skills.

Averages

An **average**, or **mean**, simplifies a set of numbers into a single number that *approximates* the value of the set.

Example: Find the average of the following set of numbers: 5, 4, 7, and 8.

Step 1: Find the sum.

$$5 + 4 + 7 + 8 = 24$$

Step 2: Divide the sum by the number of numbers in your set. Because there are four numbers in this example, divide the sum by 4.

$$\frac{24}{4} = 6$$

The average, or mean, is **6.**

Ratios

A **ratio** is a comparison between numbers, and it is usually written as a fraction.

Example: Find the ratio of thermometers to students if you have 36 thermometers and 48 students in your class.

Step 1: Make the ratio.

$$\frac{36 \text{ thermometers}}{48 \text{ students}}$$

Step 2: Reduce the fraction to its simplest form.

$$\frac{36}{48} = \frac{36 \div 12}{48 \div 12} = \frac{3}{4}$$

The ratio of thermometers to students is **3 to 4,** or $\frac{3}{4}$. The ratio may also be written in the form 3:4.

Proportions

A **proportion** is an equation that states that two ratios are equal.

$$\frac{3}{1} = \frac{12}{4}$$

To solve a proportion, first multiply across the equal sign. This is called *cross-multiplication.* If you know three of the quantities in a proportion, you can use cross-multiplication to find the fourth.

Example: Imagine that you are making a scale model of the solar system for your science project. The diameter of Jupiter is 11.2 times the diameter of the Earth. If you are using a plastic-foam ball that has a diameter of 2 cm to represent the Earth, what must the diameter of the ball representing Jupiter be?

$$\frac{11.2}{1} = \frac{x}{2 \text{ cm}}$$

Step 1: Cross-multiply.

$$\frac{11.2}{1} \diagdown\!\!\!\!\diagup \frac{x}{2}$$

$$11.2 \times 2 = x \times 1$$

Step 2: Multiply.

$$22.4 = x \times 1$$

Step 3: Isolate the variable by dividing both sides by 1.

$$x = \frac{22.4}{1}$$

$$x = 22.4 \text{ cm}$$

You will need to use a ball that has a diameter of **22.4** cm to represent Jupiter.

Percentages

A **percentage** is a ratio of a given number to 100.

> **Example:** What is 85% of 40?

Step 1: Rewrite the percentage by moving the decimal point two places to the left.

0.85

Step 2: Multiply the decimal by the number that you are calculating the percentage of.

0.85 × 40 = 34

85% of 40 is **34.**

Decimals

To **add** or **subtract decimals,** line up the digits vertically so that the decimal points line up. Then, add or subtract the columns from right to left. Carry or borrow numbers as necessary.

> **Example:** Add the following numbers: 3.1415 and 2.96.

Step 1: Line up the digits vertically so that the decimal points line up.

$$\begin{array}{r} 3.1415 \\ + \ 2.96 \\ \hline \end{array}$$

Step 2: Add the columns from right to left, and carry when necessary.

$$\begin{array}{r} \overset{1 \ \ 1}{3.1415} \\ + \ 2.96 \\ \hline 6.1015 \end{array}$$

The sum is **6.1015.**

Fractions

Numbers tell you how many; **fractions** tell you *how much of a whole*.

> **Example:** Your class has 24 plants. Your teacher instructs you to put 5 plants in a shady spot. What fraction of the plants in your class will you put in a shady spot?

Step 1: In the denominator, write the total number of parts in the whole.

$$\frac{?}{24}$$

Step 2: In the numerator, write the number of parts of the whole that are being considered.

$$\frac{5}{24}$$

So, $\frac{5}{24}$ of the plants will be in the shade.

Reducing Fractions

It is usually best to express a fraction in its simplest form. Expressing a fraction in its simplest form is called *reducing* a fraction.

> **Example:** Reduce the fraction $\frac{30}{45}$ to its simplest form.

Step 1: Find the largest whole number that will divide evenly into both the numerator and denominator. This number is called the *greatest common factor* (GCF).

Factors of the numerator 30:
1, 2, 3, 5, 6, 10, **15,** 30

Factors of the denominator 45:
1, 3, 5, 9, **15,** 45

Step 2: Divide both the numerator and the denominator by the GCF, which in this case is 15.

$$\frac{30}{45} = \frac{30 \div 15}{45 \div 15} = \frac{2}{3}$$

Thus, $\frac{30}{45}$ reduced to its simplest form is $\frac{2}{3}$.

Adding and Subtracting Fractions

To **add** or **subtract fractions** that have the **same denominator,** simply add or subtract the numerators.

Examples:

$$\frac{3}{5} + \frac{1}{5} = ? \text{ and } \frac{3}{4} - \frac{1}{4} = ?$$

Step 1: Add or subtract the numerators.

$$\frac{3}{5} + \frac{1}{5} = \frac{4}{} \text{ and } \frac{3}{4} - \frac{1}{4} = \frac{2}{}$$

Step 2: Write the sum or difference over the denominator.

$$\frac{3}{5} + \frac{1}{5} = \frac{4}{5} \text{ and } \frac{3}{4} - \frac{1}{4} = \frac{2}{4}$$

Step 3: If necessary, reduce the fraction to its simplest form.

$\frac{4}{5}$ cannot be reduced, and $\frac{2}{4} = \frac{1}{2}$.

To **add** or **subtract fractions** that have **different denominators,** first find the least common denominator (LCD).

Examples:

$$\frac{1}{2} + \frac{1}{6} = ? \text{ and } \frac{3}{4} - \frac{2}{3} = ?$$

Step 1: Write the equivalent fractions that have a common denominator.

$$\frac{3}{6} + \frac{1}{6} = ? \text{ and } \frac{9}{12} - \frac{8}{12} = ?$$

Step 2: Add or subtract the fractions.

$$\frac{3}{6} + \frac{1}{6} = \frac{4}{6} \text{ and } \frac{9}{12} - \frac{8}{12} = \frac{1}{12}$$

Step 3: If necessary, reduce the fraction to its simplest form.

The fraction $\frac{4}{6} = \frac{2}{3}$, and $\frac{1}{12}$ cannot be reduced.

Multiplying Fractions

To **multiply fractions,** multiply the numerators and the denominators together, and then reduce the fraction to its simplest form.

Example:

$$\frac{5}{9} \times \frac{7}{10} = ?$$

Step 1: Multiply the numerators and denominators.

$$\frac{5}{9} \times \frac{7}{10} = \frac{5 \times 7}{9 \times 10} = \frac{35}{90}$$

Step 2: Reduce the fraction.

$$\frac{35}{90} = \frac{35 \div 5}{90 \div 5} = \frac{7}{18}$$

Dividing Fractions

To **divide fractions,** first rewrite the divisor (the number you divide by) upside down. This number is called the *reciprocal* of the divisor. Then multiply and reduce if necessary.

Example:

$$\frac{5}{8} \div \frac{3}{2} = ?$$

Step 1: Rewrite the divisor as its reciprocal.

$$\frac{3}{2} \rightarrow \frac{2}{3}$$

Step 2: Multiply the fractions.

$$\frac{5}{8} \times \frac{2}{3} = \frac{5 \times 2}{8 \times 3} = \frac{10}{24}$$

Step 3: Reduce the fraction.

$$\frac{10}{24} = \frac{10 \div 2}{24 \div 2} = \frac{5}{12}$$

Appendix

Scientific Notation

Scientific notation is a short way of representing very large and very small numbers without writing all of the place-holding zeros.

> **Example:** Write 653,000,000 in scientific notation.

Step 1: Write the number without the place-holding zeros.

$$653$$

Step 2: Place the decimal point after the first digit.

$$6.53$$

Step 3: Find the exponent by counting the number of places that you moved the decimal point.

$$6.53000000$$

The decimal point was moved eight places to the left. Therefore, the exponent of 10 is positive 8. If you had moved the decimal point to the right, the exponent would be negative.

Step 4: Write the number in scientific notation.

$$\mathbf{6.53 \times 10^8}$$

Area

Area is the number of square units needed to cover the surface of an object.

Formulas:

$area\ of\ a\ square = side \times side$
$area\ of\ a\ rectangle = length \times width$
$area\ of\ a\ triangle = \frac{1}{2} \times base \times height$

Examples: Find the areas.

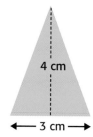

Triangle

$area = \frac{1}{2} \times base \times height$
$area = \frac{1}{2} \times 3\ cm \times 4\ cm$
$area = \mathbf{6\ cm^2}$

4 cm   *3 cm*

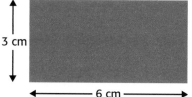

Rectangle

$area = length \times width$
$area = 6\ cm \times 3\ cm$
$area = \mathbf{18\ cm^2}$

3 cm   *6 cm*

Square

$area = side \times side$
$area = 3\ cm \times 3\ cm$
$area = \mathbf{9\ cm^2}$

3 cm   *3 cm*

Volume

Volume is the amount of space that something occupies.

Formulas:

$volume\ of\ a\ cube =$
$side \times side \times side$

$volume\ of\ a\ prism =$
$area\ of\ base \times height$

Examples:

Find the volume of the solids.

Cube

$volume = side \times side \times side$
$volume = 4\ cm \times 4\ cm \times 4\ cm$
$volume = \mathbf{64\ cm^3}$

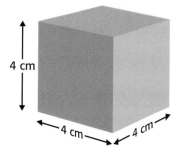

4 cm   *4 cm*   *4 cm*

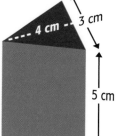

4 cm   *3 cm*   *5 cm*

Prism

$volume = area\ of\ base \times height$
$volume = (area\ of\ triangle) \times height$
$volume = (\frac{1}{2} \times 3\ cm \times 4\ cm) \times 5\ cm$
$volume = 6\ cm^2 \times 5\ cm$
$volume = \mathbf{30\ cm^3}$

Glossary

A

area a measure of the size of a surface or a region (23, 54)

astronomy the study of the universe (39)

C

compound light microscope an instrument that magnifies small objects so that they can be seen easily by using two or more lenses (21)

controlled experiment an experiment that tests only one factor at a time by using a comparison of a control group with an experimental group (12)

D

data any pieces of information acquired through observation or experimentation (77)

density the ratio of the mass of a substance to the volume of the substance (55, 86)

E

electron microscope a microscope that focuses a beam of electrons to magnify objects (21)

G

geology the study of the origin, history, and structure of the Earth and the processes that shape the Earth (36)

H

hypothesis (hie PAHTH uh sis) an explanation that is based on prior scientific research or observations and that can be tested (10, 44, 75)

L

law a summary of many experimental results and observations; a law tells how things work (18, 83)

life science the study of living things (4)

M

mass a measure of the amount of matter in an object (24, 54, 85)

meteorology the scientific study of the Earth's atmosphere, especially in relation to weather and climate (38)

meter the basic unit of length in the SI (symbol, m) (53)

model a pattern, plan, representation, or description designed to show the structure or workings of an object, system, or concept (16, 48, 80)

O

observation the process of obtaining information by using the senses (73)

oceanography the scientific study of the sea (37)

P

physical science the scientific study of nonliving matter (67)

S

science the knowledge obtained by observing natural events and conditions in order to discover facts and formulate laws or principles that can be verified or tested (66)

scientific methods a series of steps followed to solve problems (8, 43, 72)

T

technology the application of science for practical purposes; the use of tools, machines, materials, and processes to meet human needs (20)

temperature a measure of how hot (or cold) something is; specifically, a measure of the average kinetic energy of the particles in an object (24, 54, 86)

theory an explanation that ties together many hypotheses and observations (18, 50, 82)

V

variable a factor that changes in an experiment in order to test a hypothesis (12)

volume a measure of the size of a body or region in three-dimensional space (23, 53, 86)

Spanish Glossary

A

area/área una medida del tamaño de una superficie o región (23, 54)

astronomy/astronomía el estudio del universo (39)

C

compound light microscope/microcopio óptico compuesto un instrumento que magnifica objetos pequeños de modo que se puedan ver fácilmente usando dos o más lentes (21)

controlled experiment/experimento controlado un experimento que prueba sólo un factor a la vez, comparando un grupo de control con un grupo experimental (12)

D

data/datos cualquier parte de la información que se adquiere por medio de la observación o experimentación (77)

density/densidad la relación entre la masa de una substancia y su volumen (55, 86)

E

electron microscope/microscopio electrónico microscopio que enfoca un haz de electrones para aumentar la imagen de los objetos (21)

G

geology/geología el estudio del origen, historia y estructura del planeta Tierra y los procesos que le dan forma (36)

H

hypothesis/hipótesis una explicación que se basa en observaciones o investigaciones científicas previas y que se puede probar (10, 44, 75)

L

law/ley un resumen de muchos resultados y observaciones experimentales; una ley dice cómo funcionan las cosas (18, 83)

life science/ciencias de la vida el estudio de los seres vivos (4)

M

mass/masa una medida de la cantidad de materia que tiene un objeto (24, 54, 85)

meteorology/meteorología el estudio científico de la atmósfera de la Tierra, sobre todo en lo que se relaciona al tiempo y al clima (38, 51, 70)

meter/metro la unidad fundamental de longitud en el sistema internacional de unidades (símbolo: m) (53)

model/modelo un diseño, plan, representación o descripción cuyo objetivo es mostrar la estructura o funcionamiento de un objeto, sistema o concepto (16, 48, 80)

O

observation/observación el proceso de obtener información por medio de los sentidos (73)

oceanography/oceanografía el estudio científico del mar (37)

P

physical science/ciencias físicas el estudio científico de la materia sin vida (67)

S

science/ciencia el conocimiento que se obtiene por medio de la observación natural de acontecimientos y condiciones con el fin de descubrir hechos y formular leyes o principios que puedan ser verificados o probados (66)

scientific methods/métodos científicos una serie de pasos que se siguen para solucionar problemas (8, 43, 72)

T

technology/tecnología la aplicación de la ciencia con fines prácticos; el uso de herramientas, máquinas, materiales y procesos para satisfacer las necesidades de los seres humanos (20)

temperature/temperatura una medida de qué tan caliente (o frío) está algo; específicamente, una medida de la energía cinética promedio de las partículas de un objeto (24, 54, 86)

theory/teoría una explicación que relaciona muchas hipótesis y observaciones (18, 50, 82)

V

variable/variable un factor que se modifica en un experimento con el fin de probar una hipótesis (12)

volume/volumen una medida del tamaño de un cuerpo o región en un espacio de tres dimensiones (23, 53, 86)

Index

Boldface page numbers refer to illustrative material, such as figures, tables, margin elements, photographs, and illustrations.

Index

Index

Index

predictions from hypotheses, 11, **11,** 113
summary of steps in, **43, 72**
testing hypotheses, 44–45, 76–77, **76**
theories and laws, 18–19
scientific models, 16–19, **16, 17, 18,** 48–51, **49**
choosing, 50, **50**
climate models, 51, **51**
conceptual, 49
of the Earth, 50, **50**
mathematical, 49, **49,** 51, **51**
physical, 48, **48**
scientific notation, 122
scientific theories, 18–19
seismologists, 36
Seismosaurus hallorum, 42, **42, 46, 47**
SEM (scanning electron microscopes), 21, **21**
sharp object symbol, **25**
Siberian tigers, 7, **7**
SI units, 22, **22,** 52–55, **52, 53, 54,** 85–86, **85, 111**
slopes of graphs, 117–118
space shuttle, 80, **80**
spider map instructions (Graphic Organizer), 109, **109**
square, area of, 122
stars
  number of, 39, **39**
storms
  hurricanes, 38, **38**
  tornadoes, 38, **38,** 62
Sue (fossil dinosaur), 63
swimming, forces in, 75, **75**
symbols, safety, **55**

T

table fold instructions (FoldNote), 108, **108**
technology, 20, **20**
telescopes

importance to astronomy, 39, **39**
optical, 9
radio, 39, **39**
TEM (transmission electron microscopes), 21, **21**
temperature, 24, **24,** 54, **54,** 86, **86.** *See also* body temperature
body, 54, **54**
labs on, **37**
thermometers, 24, **24**
units of, **22,** 24, **24, 52,** 112, **112**
temperature scales, 24, **24,** 112, **112**
"The Time Machine," 94
theories, scientific, 18, **18,** 50, **50,** 82, **82**
thermal pollution, **69**
thermometers
  labs on, **37**
  temperature scales on, **52,** 54, **54**
time travel, 94
Tonegawa, Susumu, **6**
tons, metric, 85
tools, 20–25
  in life sciences, 9, **9**
  for measurement, **24**
tornadoes, 62, 70, **70**
  tornado chasers, 38, **38**
transmission electron microscopes (TEM), 21, **21**
Trefry, John, 37
triangle, area of, 122
Triantafyllou, Michael, 72–78, **73**
tri-fold instructions (FoldNote), 108, **108**
triple-beam balances, 115, **115**
tube worms, 37, **37**
two-panel flip chart instructions (FoldNote), 108, **108**
Tyrannosaurus rex, 63

U

ultraviolet (UV) radiation
  effect on frogs, 9–14, **10, 11, 12, 13**

underwater caves, 36, **36**
units
  of length, **52,** 53, 85, **85**
  of mass, **52,** 54, 85, **85**
  of temperature, **52**
  of volume, **52,** 53, **85,** 86
units of measurement, 22, **22,** 111, **111**
UV (ultraviolet) radiation
  effect on frogs, 9–14, **10, 11, 12, 13**

V

variables, 12, **12,** 44
vents
  hydrothermal, 37, **37**
Vermeij, Geerat, **5**
volcanoes
  physical models of, 48, **48**
volcanologists, 36
volume, 23, **23,** 53, **53,** 86, **86**
  formulas for, 122
  lab on, 88–89
  measuring, 115
  of liquids, **85,** 88–89
  units of, **22,** 23, **52,** 53, **85,** 86, **111**

W

water
  boiling point of, **54**
  freezing point of, **54**
  freezing and boiling points, **24,** 112
waterspouts, 62
weather
  tornadoes, **38,** 62
weather forecasting, 70, 81, **81**
weighing procedures, 115
Wells, H. G., 94
Williams-Byrd, Julie, 95
world population growth, **49**

Credits

Abbreviations used: (t) top, (c) center, (b) bottom, (l) left, (r) right, (bkgd) background

PHOTOGRAPHY

Cover and Title Page Mark Conlin/USFWS/FWC/Seapics.com

Table of Contents iv (tl), Chip Simmons/Discover Channel; (bl), Dr. Howard B. Bluestein; v, Gunnar Kullenberg/Stock Connection/Picture Quest, vi, vii (both), Sam Dudgeon/HRW.

Chapter One 2-3 Craig Line/AP/Wide World Photos; 4 (b), Peter Van Steen/HRW; 5 (l), NASA; 5 (c), Gerry Gropp; 5 (r), Chip Simmons/Discover Channel; 6 (t), Hank Morgan/Photo Researchers, Inc.; 6 (b), Mark Lennihan/AP/Wide World Photos; 7 © National Geographic Image Collection/Dale Miquelle; 9 (tr), Peter Van Steen/HRW; 9 (b), Sam Dudgeon/HRW; 10 Sam Dudgeon/HRW; 12 John Mitchell/Photo Researchers; 14 (b), Sam Dudgeon/HRW; 15 John Mitchell/Photo Researchers; 16 © Royalty-Free/CORBIS; 18 Art by Christopher Sloan/Photograph by Mark Thiessen both National Geographic Image Collection/© National Geographic Image Collection; 20 (bl), Alfred Pasieka/Photo Researchers; 20 (bl), Howard Sochurek/The Stock Market; 21 (tl), CENCO; 21 (bl), Robert Brons/Biological Photo Service; 21 (tc), Sinclair Stammers/Science Photo Library/Photo Researchers; 21 (tr), RJ Lee Instruments Limited; 21 (bc), Microworks/Phototake; 21 (br), Visuals Unlimited/Karl Aufderheide; 22 (t), Victoria Smith/HRW; 22 (bc), Victoria Smith/HRW; 22 (b), Victoria Smith/HRW; 22 (tc), Sam Dudgeon/HRW; 23 (bl), Peter Van Steen/HRW; 23 (br), Peter Van Steen/HRW; 25 (b), Dr. Jeremy Burgess/Science Photo Library/Photo Researchers, Inc.; 26 Sam Dudgeon/HRW; 27 Sam Dudgeon/HRW; 28 (b), Peter Van Steen/HRW; 28 (t), John Mitchell/Photo Researchers; 32 (l), Craig Fugii/©1988 The Seattle Times; 33 (r), NASA; 33 (l), NASA.

Chapter Two 34-35, © Louie Psihoyos/psihoyos.com; 36, James W. Rozzi; 37, Woods Hole Oceanographic Institute; 38 (t), Marit Jentof-Nilsen and Fritz Hasler/NASA Goddard Laboratory for Atmospheres; 38 (b), Howard B. Bluestein; 39, Jean Miele/Corbis Stock Market; 40 (bl, br), Andy Christiansen/HRW; 40, Mark Howard/Westfall Eco Images; 41, Annie Griffiths Belt/CORBIS; 44, Dr. David Gillette; 47, Paul Fraughton/HRW; 48 (r), Jim Sugar Photography/CORBIS; 48 (l), Sam Dudgeon/HRW; 50 (l), AKG Photo, London; 50 (r), Image Copyright ©2005 PhotoDisc, Inc.; 51, Andy Newman/AP/Wide World Photos; 53 (l, r), Peter Van Steen/HRW; 56, Victoria Smith/HRW; 59 (b), Andy Christiansen/HRW; 59 (l), Peter Van Steen/HRW; 62 (tl), Scripps Institution of Oceanography; 62 (tr), The Stuart News, Carl Rivenbark/AP/Wide World Photos; 63, AFP/CORBIS; 63 (b), AFP/CORBIS.

Chapter Three 64-65 (all), © Kevin Schafer/Getty Images; 66 (bl), Peter Van Steen/HRW; 67 (tl), © Jeff Hunter/Getty Images; 67 (br), Roy Ooms/Masterfile; 68 (bl), © Lawrence Livermore National Laboratory/Photo Researchers, Inc.; 69 (tl), © Gunnar Kullenberg/Stock Connection/PNI; 70 (tl), Howard B. Bluestein; 70 (bc, br), Andy Christiansen/HRW; 71 (tr), John Langford/HRW; 73 (br), Stephen Maclone/HRW; 73 (bl), Barry Chin/Boston Globe; 76 (b), Donna Coveney/MIT News; 80 (c), Digital Image copyright © 2005 PhotoDisc; 80 (bl), Peter Van Steen/HRW; 81 (bl), © JULIAN BAUM/Photo Researchers, Inc.; 82 (tl), Victoria Smith/HRW; 83 (all), Victoria Smith/HRW; 84 (all), Victoria Smith/HRW; 88 (b), Sam Dudgeon/HRW; 89 (bl), Digital Image copyright © 2005 PhotoDisc; 90 (bl), Sam Dudgeon/HRW; 90 (tl), Victoria Smith/HRW; 91 (c), John Langford/HRW; 91 (cr), Victoria Smith/HRW; 94 (tr), © Layne Kennedy/CORBIS; 95 (all), Louis Fronkier/Art Louis Photographics/HRW.

Study Skills 106 (br), 107 (br), Victoria Smith; 115 (tr), Peter Van Steen/HRW; 115 (br), Sam Dudgeon/HRW.

Copyright © by Holt, Rinehart and Winston. All rights reserved.

Credits